PROTECT Your Garden

PROTECT YOUR GARDEN
Copyright © 2013 Ed Rosenthal
Published by Quick American
A Division of Quick Trading Co.
Oakland, CA

ISBN: 978-0-932551-19-1
eISBN: 978-1-936807-16-1

Printed in North America
First Printing

Project Director: Jane Klein
Editors: Eva Saelens, David Sweet, Jennifer L. Carson
Photo Editor: Hera Lee
Design and Production: Alvaro Villanueva
Cover Photo: Christian Peacock

Library of Congress Cataloging-in-Publication Data

Rosenthal, Ed.

Protect your garden / by Ed Rosenthal.

    p. cm.

Includes bibliographical references.

  ISBN 978-0-932551-19-1

1. Plants, Protection of. 2. Agricultural pests—Control. 3. Phytopathogenic microorganisms—
Control. 4. Plant nutrients. I. Title.

SB950.R67 2013

632'.9—dc23

2012050015

PROTECT Your Garden

by Ed Rosenthal

Acknowledgements

This book is the result of a great team effort. I'd like to thank everyone who was involved in this project: Jane Klein, Nick Rosenthal, Kathy Imbriani, Hera Lee, Alvaro Villanueva, Eva Saelens, Jason Schulz, Kristine Borghino, Christian Peacock and to the many people who lent their photos to the project.

———————

Nothing's for certain
It can always go wrong
Come in when it's raining
Go on out when it's gone
We could have us a high time
Living the good life
Well I know

Lyrics: Robert Hunter and Jerry Garcia,
"High Time," Courtesy of the Grateful Dead

MOST COMMON PROBLEMS
PESTS

Aphids

Cucumber Beetles

Japanese Beetles

Tobacco Budworms

Tomato Hornworms

Earwigs

Flea Beetles

Fungus Gnats

Lace Bugs

Mealybugs

Moles

Spider Mites

Squash Bugs

Thrips

Whiteflies

MOST COMMON PROBLEMS
DISEASES

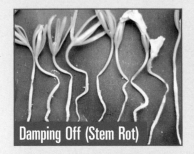
Damping Off (Stem Rot)

Gray Mold and Brown Mold

Powdery Mildew

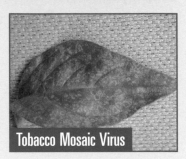
Tobacco Mosaic Virus

MOST COMMON PROBLEMS
NUTRIENTS

Calcium

Iron

Magnesium

Nitrogen

Phosphorus

Potassium

CONTENTS

NUTRIENTS

ENVIRONMENTAL STRESSES

CONTROLS: ECO-FRIENDLY SOLUTIONS

CONTROLS: BENEFICIAL BIOLOGICALS

PROTECT YOUR GARDEN

This book is a troubleshooting guide for indoor and outdoor gardeners. It is meant to take you over the bumps and help you solve garden problems. Using the information provided here, you will be able to enjoy your garden, free from troublesome pests or diseases. At the same time you will have no need to worry that the cure may be worse than the disease.

This is a family- and pet-friendly book. Rather than use dangerous synthetics, with unplanned and unwanted side effects, we use barriers, biological controls, and sensible green pesticides. These give you the freedom to be in the garden, to enjoy it, and to share its bounty without worry.

Protect Your Garden provides the tools you need to grow healthy plants, whether they are ornamentals or vegetables, annuals or perennials. It is easy to grow a garden without toxic chemicals and poisons, and it is far more satisfying. What fun is it to be in the sublime surroundings of a garden when you realize that the chemicals used to maintain it can be as deadly to you as to their target?

The methods discussed here are very effective but don't have the "seriousness" of the toxins that are on the market. Instead, you will feel the harmony developing with this little piece of nature.

Section 1 covers pests that are most likely to be found in gardens, indoors and out, small and large. Each chapter begins with a description of the pest, followed by its effect on the plant. Then a variety of preventative and problem-solving techniques are presented.

Section 2 identifies diseases that are likely to attack plants and explains how to prevent or control these diseases.

Section 3 provides information about the nutrients that plants need and how to identify

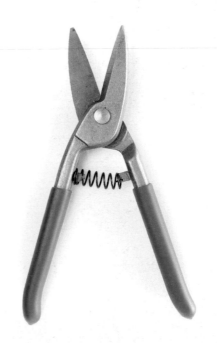

deficiencies. It also includes a guide to pH management. Fertilization is important because plants need proper nutrition to grow properly.

Section 4 reviews environmental stresses that can impact garden plants and how to alleviate them. Sometimes plants fail to thrive because of environmental conditions. This section helps you analyze a plant's situation and show you how to improve it.

Section 5 is a guide to the controls referenced in the preceding sections, with suggested commercial brands. This section is divided into two parts: Controls, which includes pesticides and fungicides, and Beneficial Biologicals, which includes different organisms used to control diseases

This book shows you how to protect your garden safely. The harvest from plants, whether flowers or food, will be safe for you and yours to be around, touch, and eat, without the fear of chemical residues tainting your experience. The solutions in this book are friendly to humans, pets, and the planet, and all are extremely effective.

1. PESTS

No matter whether your garden is inside or out, large or small, pests in the garden are among the most annoying and difficult problems. No matter the growing method, pests can infect the garden. Outdoors, pests walk, fly, crawl, or float into the garden. Pests travel indoors on clothes and pets.

This section provides information to help you to recognize and eradicate pests in your garden. Like the entries in the other sections, the pests are listed in alphabetical order, but the ones that are most likely to attack your plants are aphids, caterpillars, fungus gnats, mealybugs, scale, spider mites, and whiteflies. A description

At the end of each chapter is a summary of solutions specific to the pest. No store can carry all brands so familiarize yourself with the active ingredients listed. For more information on when and how to use the controls look them up in Section 5, which also includes referrals to commercially available products.

For many problems redundancy—using more than one solution—is good because it creates multiple modes of action.

of each pest is provided so that you can identify the pest and the damage that it does. Preventative and control methods are listed to keep the pest away from plants and to get rid of an infestation.

GOOD PRACTICES

Here are generally good practices to employ to prevent outbreaks in both your indoor and outdoor gardens.

- ➡ **Infected plants:** Quarantine all new plants for 10 to 15 days in a separate area. Give them a prophylactic treatment with a pesticide.

- ➡ **Infected planting mix:** Use new pasteurized or sterile planting mix, especially for seedlings and transplants. If you suspect contaminants in incoming soil on plants, remove the soil, wash the roots, and replant, or just take cuttings and start new plants from them.

- ➡ **Clothing:** Don't go into your indoor garden after being outside near plants. You could be transporting a pest. Wear a hat or a head covering, wash yourself, and change clothes. Put a welcome mat in a tray filled with salt water so that the bottoms of your shoes are wetted with the salt solution, killing pests.

➤ **Pets:** Dogs and cats may carry in pests as well as diseases. Keep your pets out of the indoor garden. You may wish to keep them out of the outdoor garden, too.

➤ **Air and ventilation system, open windows, and row covers:** To kill airborne pests and pathogens, put filters on the intake vents of your house and place fixtures from a UVC light kit in the air and ventilation system. UVC light kits emit UVC light, which is lethal to most life forms, including fungal spores and bacteria. They are sometimes seen in restaurants as a small fluorescent tube emitting purple light, usually placed in an inconspicuous space. They are available online. Additionally, screen open windows.

Outdoors, use row covers to keep pests from delicate leaves and to protect plants from flying invaders. They can also be used to keep insects in the garden. For instance, ladybeetles ordinarily disperse, but not when they are under the row covers.

Ants

How common are these pests?

Ants are abundant all over the world, both indoors and out.

What do they look like?

The ant body, like that of all insects, is made up of three segments: the head, thorax, and abdomen. Ants differ from other insects in two ways: their antennae are elbowed and the first segment of their abdomen is strongly constricted, giving them a "waist." The six legs are attached to the thorax. The eyes, jaws, and antennae are all found on the head. Commonly seen ants range in size from less than 0.1 inch up to 1 inch.

What kinds of plants do they attack?

Most species of ants encountered in the garden do not attack plants. They are likely to be herding aphids or mealybugs for their sugary "honeydew" or preying on other small organisms. Some species of ants do feed on seeds, seedlings, or soft plant tissue. These include leafcutter ants, which can defoliate shrubs. They use the leaf fragments as a medium to grow a specific fungus that they use for food.

Ant tending milkweed aphids

Where are they found on the plant?

Ants are found on stems of climbing plants or on the leaves, tending their colonies of aphids or mealybugs. They

Carpenter ant

Fire ant

Argentine ant

also tunnel around the roots of plants or live in the planting media.

What do they do to the plant?

Ant-managed mealybug and aphid colonies suck vital plant juices to produce honeydew that ants love. Its presence promotes the growth of sooty mold, a black, nonpathogenic fungus that can block light from reaching the leaf surface. Ants that burrow around plant roots damage root tissue and root hairs.

General Discussion

Ants are attracted to plants that already have a population of aphids, mealybugs, scale, or whiteflies. Once ants find sufficient numbers of their honeydew "cows," they move them to new feeding areas, spreading them throughout the plant. A steady stream of ants crawls up and down the stems of affected plants and then across the growing medium. Once they exhaust a feeding area, ants move their herd to other plants and other feeding areas.

Ants are social and live in colonies of queens and different castes of worker ants. Some species have one queen while others have several. The life cycle begins when the queen lays an egg. After hatching, the young ant passes through the larval, pupal, and adult stages. New colonies are produced when a new queen, along with several carefully selected males, leaves the nest to find a new location.

Ants regulate their reproductive rate based on conditions in the colony and the outside environment. This is accomplished partly by regulating the length of the pupal stage and partly by selecting certain larvae or pupae for transformation into new queens and drones (worker ants). If suitable weather and ample food and water are present, reproduction increases. Conversely, when the nest is stressed by inclement weather or scarcity of food and water, reproduction slows.

In permanent nests, 90 percent of the ants work inside. The significance of this is that destroying only the ants that you can see is not enough to protect your garden. Lost workers are quickly replaced. Instead, entire colonies must be destroyed.

Exclusion and Prevention

► **Moats:** Place potted plants on blocks of wood or Styrofoam in wide trays of water to act as "moats." Ants can't swim and will not be able to reach the plant.

► **Herbal repellants:** Deter ants with repellant spices: bay leaves, cinnamon, and cloves. Pour the ground spice around the perimeter of the growing area to make an herbal barrier that ants won't cross. A tea brewed from these herbs and sprayed on plants also deters ants.

► Create a perimeter of boric acid powder around the garden or between plants. Ants may cross the barrier, but their bodies will be coated with the acidic powder that slowly kills them. They live long enough to carry the poison back to their nest to share with their nest mates.

► Diatomaceous earth, composed of the silica skeletons of long-dead sea animals, is available as a finely ground powder for use as a barrier. Its sharp points pierce the ant's exoskeleton, causing death. Diatomaceous earth must remain dry because it is ineffective when wet.

► Coat tree and shrub stems with Tanglefoot or other sticky substances such as

Leafcutter ant making the cut

Leafcutter ants carrying leaf cuts to the nest

Leafcutter ants tending to food fungus growing on leaves

Ants herding aphids

heavy oil or petroleum jelly to keep ants from climbing stems. Sticky cards, paper, or flypaper also stop the trekking.

CONTROLS

➤ Ant stakes and ant baits use minute amounts of poison to target and kill ants. They are very safe. Ants carry these tiny bits of poison back to their nests.

➤ **Boric acid bait:** Ants are interested in grease, oils, or sugar. Mix sweets or fats with boric acid to make a toxic ant lunch.

➤ **Herbal solutions:** Use cayenne, cinnamon, citrus rind, clove, "Italian" seasoning, lemongrass, mint, rosemary, and/or thyme as powder, tea, or oil, alone or in combination, to kill and repel ants from planting mediums. Sprinkling these spices on the surface of the soil or watering with a tea brewed from them is both toxic and repellant.

Pyrethrum

Spinosad

BENEFICIAL BIOLOGICALS

ANTIdote

Similarities Between Ants and Humans

1) Both live in complex social communities.

2) The communities are cooperative.

3) Both have individuals who will sacrifice themselves for the benefit of the community.

4) Both herd other organisms.

5) Both farm.

6) Both hunt in groups.

7) Both have workers to give the young extended care.

8) Both practice territoriality.

9) Both have forms of slavery.

10) Both have wars between tribes.

11) Both have colonized territories far beyond their native lands.

12) Both have the biggest brains in proportion to body for their phyla.

13) Both have complex modes of communication.

14) Both teach tasks to young members. (Teaching entails performing tasks slowly or in an exaggerated style and paying attention to the student's learning pace.)

15) Both differentiate jobs and assign tasks to specific groups or classes.

16) Individuals change tasks as they mature.

17) They are both capable of complex architecture for homes and workplaces.

18) They both use tools.

19) They both air condition their homes.

20) Each has only a small proportion of members who produce or obtain food for the entire group.

21) Individual ants cannot survive alone. They need the community.

Differences Between Ants and Termites

Ants	Termites
Two pairs of wings.	Wings all the same size.
Veins in wings easily seen.	Veins present but not noticeable and often reduced to veinlike wrinkles.
Wings do not break off easily.	Wings break off easily. Broken wings are plentiful in swarming area.
Antennae are elbowed.	Antennae are straight.
Front pair of wings extend just past end of body.	Both pairs of wings twice as long as body.
Bodies have very constricted waist.	Bodies straight without waist.

Aphids

How common are these pests?

Aphids are a common pest found indoors and outdoors, all over the world.

What do they look like?

Aphids are small, soft-bodied, pear-shaped insects about ⅛ inch long. Their colors include black, brown, gray, green, red, and yellow. Their most distinguishing features are two structures that look like tail pipes, called cornicles, which extend from the end of their abdomen. Aphids are true bugs, all of the order Hemiptera. Hemiptera mouthparts have evolved into a straw-like proboscis which they use to use pierce tissue and suck the juices. Aphids suck juices, pierce leaves and stems, to suck juices from the phloem layer. Most species of aphids have winged stages. Other species cover their bodies with long, waxy or "wooly" threads to deter their natural enemies. Cast-off exoskeletons from molting larvae appear as white specks on the undersides of leaves mingled among the aphids.

Aphids have distinctive "cornicles" that look like tailpipes extending from their abdomen.

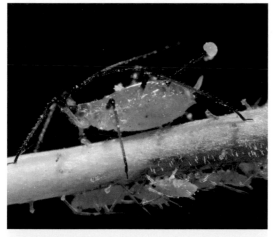

Aphid family

Aphids like to cluster in herds.

What kinds of plants do they attack?

Aphids attack a wide variety of plants ranging from annual flowers and vegetables to evergreen trees and ornamental shrubs. Some of the most common hosts are carrots, chrysanthemums, cucumbers, geraniums, hops, milkweed, nasturtiums, peppers, roses, and tomatoes.

Where are they found on the plant?

Aphids cluster on the undersides of leaf surfaces, the stem, and petioles of plants. They occur in profusion and are rarely found alone or in small numbers. Some species are very colorful and stand out on leaf surfaces, such as the red poplar leaf aphid. Others, including the green peach aphid, are nearly the color of the leaves and can be hard to see. Aphids excrete a "honeydew," a sweet, sticky liquid that becomes shiny as it dries A black, noninvasive mold called sooty mold grows on it, and ants milk the aphids for it. Aphids cast off exoskeletons when they molt.

What do they do to the plant?

Aphids pierce plant tissue and suck up the plant juices, leaving infested plant tissue curled, wilted, stunted, distorted, and discolored. Leaves appear pale and "stippled" where the plant tissue has been punctured. Sooty mold growing on abundant honeydew can become so dense that it blocks light. Aphids are also vectors for a large number of disease-causing organisms, including bacteria, viruses, and fungi, transferring them from plant to plant as they feed.

General Discussion

Aphids are prolific breeders and can overrun a garden quickly. Outdoors in temperate zones their life cycle involves several generations of asexual reproduction with one sexual egg producing one generation per year. The eggs are laid and overwinter on a perennial host plant (called the primary host) that is different from the plants on which they feed (the secondary host).

Females hatch from those eggs and bear live females from asexually produced eggs that hatch within their bodies. These offspring complete their life cycle in seven to 14 days. Each female produces about 100 babies. When they become overcrowded or the day length and the temperature indicate cooling weather, they produce winged daughters. During summer they migrate to new food sources. As fall approaches the winged aphids fly to their primary hosts and produce a generation that includes males. They mate with the females, which produce the eggs that overwinter.

Indoors and in greenhouses, protected from the elements and safe from most predators, the egg stage is eliminated completely and aphids reproduce year-round.

Exclusion and Prevention

- Row covers prevent aphids from finding and feasting on outdoor crops if installed before pests arrive. Remove row covers before pollination.

- Aphids cause the most damage when temperatures are between 65°F and 80°F, warm but not hot. Monitor plants regularly for an infestation.

- Once the leaves begin to curl and distort due to feeding, aphids are protected from insecticides, natural enemies, and other control methods.

- An abundance of ants on plants is often an indication that their favorite "cows" are present, producing honeydew.

Aphids can be different colors including black, brown, gray, green, red, and yellow.

- Aphids prefer the upwind side of gardens and also areas close to other aphid havens, such as weedy borders, so check those areas thoroughly.

- Another aphid indicator is the presence of predators such as lady beetles, lacewings, syrphid fly larvae, and the dark mummified bodies of parasitized aphids. If enough predators and parasitoids are present, the aphid population can be controlled naturally.

- Control ants. Ants herd aphids, keeping them safe from natural predators. Keeping ant populations under control helps keep aphids out of the garden.

- Avoid overfeeding—high nitrogen encourages abundant tender growth, which is attractive to aphids.

CONTROLS

Azadirachtin	Insecticidal soap
Capsaicin	Neem oil
D-limonene	Pyrethrum
Garlic	Insecticidal soap
Herbal oils (cinnamon, cloves, and coriander)	Vacuuming
	Water spray
Horticultural oil	Yellow sticky cards

BENEFICIAL BIOLOGICALS

➤ A number of organisms are natural control mechanisms for aphids. Aphid parasites such as various species of parasitic wasps are most effective when aphid numbers have not yet reached epidemic proportions. The wasp lays her eggs inside the aphids. The skin of the parasitized aphids turns brown and crusty, or "mummified." The larva feeds on the aphid, then emerges as an adult and immediately seeks a mate. Once mummies begin to appear, the population is likely to be substantially reduced within one or two weeks.

➤ Predators reduce large populations quicker. Insects such as predatory midges, green lacewings, and big-eyed bugs are among the best known. Lady beetles (*Lady Bugs*), while effective, wander away in search of more plentiful food as soon as the aphid population begins to drop.

➤ The fungus *Beauveria bassiana* is effective against a large number of insects, including aphids. It acts as a contact control and can kill whole colonies of aphids if the conditions are right. The fungus is working when dead, reddish-brown aphids with a fuzzy or shriveled texture are visible. These cadavers differ from the bloated, yellowish mummies of parasitized aphids. Collect the fungus-killed aphids and blend them (1 to 3 teaspoons of aphids per quart of water), then spray on plants to further distribute the fungus.

Aphid midges (*Aphidoletes aphidomyza*)	**Green lacewings** (*Chrysopa fufilabris*)
Beauvaria bassiana	*Lady beetles*
Big-eyed bugs (*Geocoris spp.*)	**Wasps, parasitoid** (*Aphidius colemani* and *Aphidius ervi*)

Interesting Facts

Aphids of one family, Pemphigidae, induce plants to form galls in which the aphid and her clone daughters live. Some of the daughters are born as a special soldier class. They have stronger legs and more weapon-like stylets that they use to pierce enemies. The soldier class does not mature fully but stays in an adolescent state. When the gall is attacked by a predator such as an aphid lion or a wasp, the soldiers mob the attacker, covering its body and using their stylets to puncture it to death.

Bagworms
(Thyrirlopteryx ephemeraeformis)

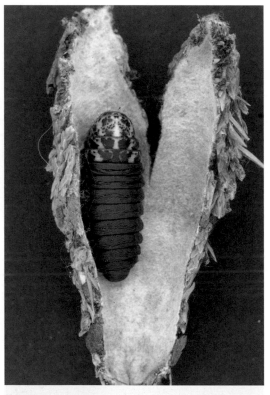

Bagworm pupa

How common are these pests?
Bagworms are the caterpillars of moths. They are found in the eastern United States, ranging from southern New England to parts of Florida, Nebraska, and Texas.

What do they look like?
Bagworms are caterpillars that live inside a "bag" they spin from silk and camouflage with sticks, needles, and bits of debris. Bags are oval shaped and are about 1½ to 2 inches long. Adult males are black, fuzzy, clear-winged moths, about 1 inch in wingspan. Females never metamorphosize into moths and instead remain in caterpillar form, spending all the stages of their lives in the bag.

What kinds of plants do they attack?
Bagworms prefer evergreen hosts such as arborvitae, cedar, cypress, false cypress, juniper pine, and spruce. They also feed on other trees, including apple, birch, buckeye, elm, locust, oak, palmetto, sycamore, and willow.

Where are they found on the plant?
The bags dangle from branches. The larvae eat the foliage and buds of their chosen hosts. Newly emerged larvae are most often found on the upper surfaces of leaves where they feed on the surface tissues. Later, they move underneath the branches and leaves and feed more generally.

What do they do to the plant?

A severe bagworm infestation seriously defoliates the host and can cause death in one or two seasons. Aside from damaging the tree, hundreds of bagworm bags hanging from the tree branches are not a pretty sight.

General Discussion

Bagworms overwinter as eggs snuggly protected inside the tough, nearly impenetrable bag. In late spring, the eggs hatch and the larvae emerge and drop down from the bag on a slender, silken thread that also acts as a sort of parachute to bear them on breezes to new feeding places. Once they find a suitable feeding location, they immediately spin their own bag. By late summer, the larvae have grown and enlarged their bag. Now full-grown caterpillars, they crawl into their bag to pupate. They

Bagworm male adult, bag, and pupal case

emerge as adults in August or September. Males leave on a mating flight in search of females that have not left their bags. Once mated, the female lays up to 1,000 eggs inside her old pupal case. Then, she emerges from her bag, drops to the ground, and dies. In most areas, there is only one generation per year, but in warmer areas, such as Florida, bagworms reproduce year-round.

Exclusion and Prevention

- ➤ Bagworms are extremely hard to control by traditional methods due to the protection of their bags. Handpicking the bags as soon as they are visible, usually in mid- to late summer, is the best method of control, though laborious. Be sure to also remove the silken band that secures the bag to the branch.

- ➤ Planting members of the aster family, such as Shasta daisies, underneath susceptible plants attracts naturally occurring enemies that prey on bagworms such as birds, mice, and parasitic wasps.

CONTROLS

Beauveria bassiana

Herbal oils

Neem oil

Pheromone traps to capture males in late summer

Pyrethrum

Spinosad

BENEFICIAL BIOLOGICALS

Bacillus thuringiensis, var. *kurstaki*

Ichneumon wasps (naturally occurring parasites attracted by plants in the Aster family)

Interesting Facts

Bagworm bags can look very different depending on the host that supports them. For instance, on a rosebush the bag is covered with bits of rose leaves or maybe even dried petals. But on a fir or arborvitae, the bag looks "prickly," covered with needles and bits of evergreen foliage. If you spot any garden plant that suddenly has "cones" where they should not be, be suspicious, collect a bag, and slice it open to see if caterpillars are lying inside.

Beetles

Asparagus Beetles
Colorado Potato Beetles
Cucumber Beetles

Flea Beetles
Japanese Beetles
Mexican Bean Beetles

The order Coleoptera, the beetles, includes more species than any other insect order, constituting between 25 and 40 percent of all insects. Only a small percentage of them have been identified and cataloged. Beetles are extremely diverse and are found in all environments, including land, soil, and freshwater and marine environments. Only a small number of the species, such as ladybugs, are predatory. The rest eat living, harvested, or decaying plant matter or fungi. For this reason, many species are pests.

Beetle species have incredible variation in life cycle, form, diet, and survival techniques. However, they do have certain morphological similarities.

All species have a hard exoskeleton. Their forewings, called elytra, are hardened and sometimes fused to form a hard protective armor over their sides and back. They lift up to release the aftwings in beetles that fly.

The exoskeletons are composed of tiny scale-like plates (sclerites) that create a flexible body shield. The rest of the general anatomy of the beetle is uniform. However, species have adapted different body parts to function in diverse ways. For instance, some have claws, others spit noxious or toxic liquids, and others have "bony" heads to engage in reproductive combat.

Beetles are endopterygotes; they undergo complete metamorphosis: egg, larva, pupa, and adult.

Beetles:
Asparagus Beetles *(Crioceris asparagi)*

Asparagus beetle

How common are these pests?
Asparagus beetles are found wherever asparagus grows. They are more likely to occur where there are many asparagus gardens in the area.

What do they look like?
Adults are ¼ inch long, bluish black with six square, pale yellow spots on their backs and red on their wing borders. Larvae are ⅛ to 1/16 inch long, light gray with a black head, and slug-like with visible heads and legs.

What kinds of plants do they attack?
Asparagus

Where are they found on the plant?
Adults chew pits into emerging spears, and larvae chew the ferns or adult growth.

What do they do to the plant?
The chewing by adults causes growth distortions such as spears curving into a shepherd's hook and ragged growing tips. Large numbers of eggs on emerging spears look unappealing. Larval chewing gives the ferns a bleached appearance, inhibits photosynthesis, and reduces the plant's ability to store food for next season. Defoliation of the ferns can also make the plant more susceptible to invasion by *Fusarium* fungi.

General Discussion

Asparagus beetles overwinter as adults in the debris and litter around asparagus beds. In spring, when the temperatures warm, these adults awaken and fly to find mates and newly emerged asparagus spears to feed on. Once they find tender new growth, they feed on the stems, and females lay black, oblong eggs, placed one end up, on the spears. Eggs hatch in about a week, and the larvae crawl upward to feed on the ferns or mature foliage of the asparagus plant. In two to three weeks, the larvae fall to the ground, burrow in, and pupate, emerging as adults in about a week. Two or three generations may be produced in a season where temperatures are warmer, but in temperate climes only one generation is produced. Newly emerged adults will feed on the ferns for the remainder of the season.

Common asparagus beetles mating

Exclusion and Prevention

➡ Clear away garden debris to eliminate overwintering areas.

➡ Use row covers to protect new spears.

➡ Harvest early and consistently throughout the production season.

➡ Vacuum or handpick eggs and adults as soon as they are visible.

➡ Brush larvae off plants onto the ground. They will be unable to climb back up on the plants and may be eaten by ants, ground beetles, or spiders.

➡ Scout for beetles in the afternoon when they are most active. Start scouting just after asparagus plants emerge and continue throughout the season.

CONTROLS

D-limonene	Neem oil
Herbal oils	Pyrethrum

BENEFICIAL BIOLOGICALS

Beauvaria bassiana	Ground beetles
Damsel bugs	Lacewing larvae

Lady beetles

Nematodes, beneficial
(*Heterorhabditis bacteriophora*)

Praying mantis

Tetrastichus asparagi, a ⅛-inch, metallic green wasp, parasitizes asparagus beetle eggs and can provide up to 70 percent control. It is a naturally occurring parasite in most asparagus gardens.

Interesting Facts

The spotted asparagus beetle *(Crioceris duo-decimpunctata)* is also a pest of asparagus and feeds on the spears and ferns as well, but it does less damage than the common asparagus beetle, preferring to lay its eggs in the berries of the asparagus plant. Orange, oblong in shape, and with 12 black spots on its back, this beetle's larvae are bright orange, far more colorful than their drab gray cousins.

Spotted asparagus beetle

Spotted asparagus beetle larva

Beetles:
Colorado Potato Beetles
(Leptinotarsa decemlineata)

How common are these pests?

The Colorado potato beetle is very common in gardens across North America except in some parts of California, eastern Canada, Florida, Nevada, and the Deep South.

What do they look like?

Adult Colorado potato beetles have a rounded, domed shell that is alternately striped with black and tan lines. They are about ⅓ inch long, and their head is reddish with several irregular black spots. Larvae are reddish and plump with a string of horizontal black dots on their sides. When full grown, they're about ½ inch long with a small, black head. Adult female beetles lay yellowish-orange eggs, end up, in masses on the undersides of leaves.

What kinds of plants do they attack?

All members of the Solanaceae family are susceptible, including belladonna, eggplant, ground cherry, henbane, horse nettle, pepper, petunia, potato, tobacco, and tomato. Colorado potato beetles rarely attack non-Solanaceae plants.

Colorado Potato Beetle

Where are they found on the plant?
Adult and larvae feed on the leaves of host plants.

What do they do to the plant?
Both adults and larvae skeletonize leaves, leaving behind only leaf veins and a trail of black flecks of excrement. Adult beetles feed on notched wounds along the leaf margins. Larvae feed in a more ragged pattern. Plants can be stripped of foliage in a short time.

Colorado Potato Beetle

General Discussion
The Colorado potato beetle is the most important potato pest worldwide. Adult beetles overwinter in the soil, buried several inches deep or in plant debris. They emerge in late spring, then fly to nearby locations to feed and mate. Females then start to lay, producing 500 to 1,000 eggs in their lifetime. Eggs hatch in four to nine days. At first larvae feed in groups, then disperse through the plant and continue to feed for another two to three weeks. Mature larvae drop from the plants to pupate in the soil. Adults emerge in about two weeks. In warmer areas, a second generation may be produced. Three generations may be produced in the most southern areas of their range. After a few weeks of late-season feeding, adults disperse, seeking out areas to overwinter. They remain dormant until the next spring.

Exclusion and Prevention

➤ Place impervious mulch such as agricultural cloth, newsprint, or woven polyethylene over the soil. The larvae that drop down to the soil to pupate are blocked and perish.

➤ Floating row covers applied immediately after planting prevent Colorado potato beetles from infesting new plants. Remove the covers for pollination (except for potatoes).

➤ Shake plants over a drop cloth in early spring to dislodge newly emerged adults. Destroy by squashing or dropping into a container of soapy water.

➤ Rotate garden planting so that members of the Solanaceae family are not planted in an infected area for at least three years. Colorado potato beetles cannot fly unless the temperature is above 70°F. In cooler locations, such as the Northeast, planting susceptible crops well away from last year's will ensure that adults will have to walk from their overwintering site to a new location.

➤ Handpick the eggs, larvae, and adults when found and squash them.

➤ Plant pollen and nectar flowers nearby to attract predators and parasites.

➤ Use companion plants that repel Colorado potato beetles. These include catnip, hemp, oak, sage, tansy, and wild potato *(Solanum chacoense)*. Plant varieties less susceptible to the beetles, such as Russet Burbank potatoes.

Colorado Potato Beetle larva

➤ Plant early-maturing varieties of potatoes, such as Sunrise, Superior, and Yukon Gold, which produce tubers in about 80 days or before beetle populations explode in late summer.

CONTROLS

Herbal oils	Pyrethrum
Neem oil	Spinosad

BENEFICIAL BIOLOGICALS

Beauveria bassiana	Nematodes, beneficial (*Steinernema feltiae, Heterorhabditis bacteriophora*)
Bacillus thuringiensis, several strains	
Lacewings	Praying mantis
Ladybugs	Wasps, parasitoid

Interesting Facts

Because of years of heavy pesticide use, Colorado potato beetles have developed resistance against most chemical controls. As a result, commercial producers have been forced to use stronger and stronger controls.

Beetles:
Cucumber Beetles

How common are these pests?

Six species of cucumber beetle exist in the United States. The eastern striped cucumber beetle *(Acalymma vittatum)* is found mostly east of the Mississippi River. The western striped cucumber beetle *(Acalymma trivittatum)* is found west of the Mississippi River. The spotted cucumber beetle *(Diabrotica undecimpunctata howardi)* is found east of the Rockies and deep into Mexico. The western spotted cucumber beetle *(Diabrotica undecimpunctata)* is found in Arizona, California, Colorado, and Oregon. The banded cucumber beetle *(Diabrotica balteata)* is found mostly in the southern United States. Corn rootworms *(Diabrotica virgifera)* are found west of the Rockies and from Ontario to North Carolina.

Adult striped cucumber beetles chewing and injuring

What do they look like?

Both the eastern and western striped cucumber beetles are about ¼ inch long, yellowish green with three black, parallel, longitudinal lines running down their wing covers, black heads, and yellow thoraxes. The spotted cucumber beetle is yellowish green with a lime-green thorax and 11 spots on its wing covers. The western spotted cucumber beetle is similar in appearance but smaller and has 12 black spots. The banded cucumber beetle is the most colorful. It is lime green with a red head, black thorax, and three horizontal greenish-blue stripes on its wing covers. Adult western corn rootworms are similar in appearance to the striped cucumber beetle except that their abdomens are yellow instead of black. Their larvae are grub-like, white with black or brown heads depending on the species.

What kinds of plants do they attack?

These beetles are frequent pests of members of the Cucurbitaceae family such as cucumbers, eggplant, melons, pumpkins, and squash. They also feed on apricots and other soft fruits, asparagus, beans, beets, corn, peas, potatoes, tomatoes, and the blossoms of other garden plants.

Western spotted cucumber beetle

Where are they found on the plant?

Adults feed on germinating seeds, seedlings, leaves, stems, blossoms, and fruit. Larvae are found in the soil and feed on the roots of plants, but they also feed on the rinds of ripening fruit that touch the ground.

What do they do to the plant?

Adults chew ragged holes in leaves, feeding heavily from the underside, and also consume whole flowers. Some species feed on germinating seeds and seedlings. Larval feeding can cause extensive damage, including plant death, wilting, and lodging. Cucumber beetles transmit bacterial wilt *(Erwinia tracheiphila)*, cucumber mosaic virus, and squash mosaic virus.

General Discussion

Cucumber beetles overwinter as adults under leaves, in debris, or in weedy areas. They emerge in spring when temperatures reach 55°F to 65°F, mate, and lay eggs around the base of host plants or in soil cracks. Egg production per female ranges from 200 to 1,500 eggs, laid in several batches over the season. Eggs hatch in five to 10 days. The larvae feed on roots for several weeks before pupating and emerging as adults. Adults live up to 60 days. One or two generations are produced each year, but more are possible depending on the length of the season.

Spotted cucumber beetle larva

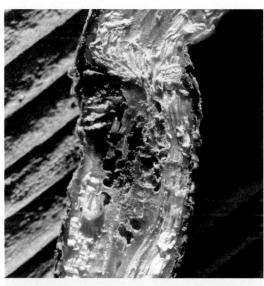
Cucumber beetle damage: extensive tunneling by larvae at plant base producing plant collapse

Exclusion and Prevention

➤ Use floating row covers.

➤ Plant fragrant catnip, marigolds, or tansy, with susceptible plants to deter adults.

➤ Cover the soil with an impermeable barrier such as heavy-duty agricultural cloth, newspaper, or rugs to prevent females from burying their eggs.

➤ Delay planting susceptible crops for a week or two to avoid the early-spring feeding of newly emerged adults.

➤ Females are attracted to the color yellow and to allspice, clove, and bay oil. Place yellow sticky cards with a cotton ball doped with any of these spice oil just above plant tops and around the garden perimeter to trap incoming females.

➤ Remove the overwintering environments of adults by clearing away old leaves and debris before they emerge in spring.

➤ Handpick or vacuum beetles off plants.

➤ Shake beetles from the plants onto tarps.

CONTROLS

Herbal oils	Pyrethrum
Neem oil	Spinosad

BENEFICIAL BIOLOGICALS

Assassin bugs	Tachinid flies
Beauvaria bassiana	Wasps, parasitoid
Nematodes, beneficial (*Steinernema feltae*, *S. riobravis*, or *Heterorhabditis bacteriophora*)	

Beetles:
Flea Beetles

How common are these pests?

Flea beetles are common throughout North America. They are pests in many gardens. They rarely cause significant damage but may affect the appearance of ornamentals and flowers.

What do they look like?

Adult beetles are brown or black, sometimes with a bronze or reddish hue, and mostly uniformly colored, although some species have stripes. They are among the smallest of the leaf beetles, about 1/10 inch long. Their hind legs are large and allow them to jump like fleas when disturbed. Larvae are legless, thin, white grubs, up to ¾ inch long, that live in the soil.

Ragwort flea beetle

What kinds of plants do they attack?

Adult flea beetles attach to cabbage and crucifers, corn, eggplant, flowers, pepper, potato, radish, tomato, turnip, many weeds, and flowering shrubs. Different species prefer different groups of plants, for instance, crucifers or nightshades (Solanaceae).

Where are they found on the plant?

Adult flea beetles feed on both the upper and lower sides of the leaves and along the stems. They prefer cotyledons, stems, foliage, and tender, young vegetable seedlings. Larvae feed on plant roots but do little damage. When disturbed, flea beetles use their powerful hind legs to jump, looking much like fleas.

What do they do to the plant?

Infested leaves have small, roundish holes that look as if they've been riddled with buck-

Ragwort flea beetle

shot. A large infestation can rapidly defoliate and kill plants.

General Discussion

Flea beetles are a subfamily of leaf beetles and are divided into two lifestyle categories: those that develop below ground and those that develop on foliage. The most common species develop beneath the soil.

Flea beetles overwinter as adults in the soil. They emerge in spring and lay eggs in cracks in the soil around the base of plants. These adults then die by early July. Eggs hatch in about a week, and the larvae stay on the leaves or crawl into the soil depending on species, and feed for two to three weeks. The leaf eaters return to the soil to pupate. The new adults emerge in two to three weeks. One to four generations are produced per year depending on locale and temperature.

Exclusion and Prevention

➤ Clean up garden debris to eliminate overwintering sites.

➤ Cultivate the top ½ inch of soil near the base of plants to expose flea beetle eggs to predators.

➤ Use floating row covers.

➤ Place barriers over soil to stop adults emerging from metamorphosis from becoming airborne.

➤ Delay planting to avoid spring flush of adults.

➤ Water during dry spells—flea beetles prefer dry conditions.

➤ Plant onion and mint as repellents.

CONTROLS

Citrus oil	Neem oil
Garlic oil	Pyrethrum
Herbal oils	Spinosad

BENEFICIAL BIOLOGICALS

Beauveria bassiana	Nematodes, beneficial
Bacillus thuringiensis var. *israelensis*	(*Heterorhabditis bacteriophora*)

Beetles:
Japanese Beetles *(Popillia japonica)*

How common are these pests?

First found in the United States in a nursery in New Jersey in 1916, the Japanese beetle is now an important pest in most eastern states and parts of southern Canada, but it is rapidly moving westward. Many trees, shrubs, vegetables, and flowers are affected. Infestations are common in the eastern United States.

What do they look like?

Adults are oval, about ½ inch long, and metallic green with bronze or copper wing covers. White bristles of hair rim the sides and back beneath the edges of the wings.

Japanese Beetle adult

Males are slightly smaller than females. Adults are often seen mating in chains hanging from plants or structures such as the eaves of houses. The larvae, white grubs found in the soil, typically lay in a C-shape when at rest. The abdomen is translucent so dark fecal matter can be seen through the skin.

What kinds of plants do they attack?

The host range of Japanese beetles is wide and varied, including more than 300 plants. They are ravenous pests of field and vegetable crops, flowers, grasses, shrubs, trees, and vines. A few plants affected include beans, corn, grapes, peas, and roses. Trees include apple, ash, birch, cherry, chestnut, elm, linden, maple, oak, plum, sycamore, and willow.

Where are they found on the plant?

Adults are found on the leaves of plants while the grubs live 4 to 8 inches underground.

Japanese Beetle skeletonizing European linden

What do they do to the plant?

Adults feed on flower petals, foliage, and fruits. Larvae feed on grass and plant roots.

Both adults and larvae do significant damage to their hosts. The adults feed in groups and eat the leaf tissue between the veins, or skeletonize the leaves. High numbers can strip a plant in short time. Larvae eat the roots of plants and grasses and can quickly devastate a golf course, park, or pasture, leaving behind irregular brown patches where the turf peels up easily.

General Discussion

Japanese beetles overwinter as nearly mature grubs that lay curled in a C-shape 4 to 8 inches beneath the soil. When spring temperatures warm the soil, the grubs rouse and resume feeding on grass roots until late spring when they burrow deeper in the soil and form a round, tamped cavity and pupate. In late June adults emerge and begin to feed and mate in large groups. Aggregation pheromones combined with the odors produced by favorite plants can attract large numbers to one place.

Japanese beetles prefer full sun and temperatures between 85°F and 95°F. Their feeding is heaviest then. Females interrupt feeding and leave the plant, burrow 1 to 3 inches into the soil, or preferably turf, and lay a few eggs. Females continue this routine every few days until each female has laid 40 to 60 eggs. Most eggs are laid by August or September. Eggs hatch, and the larvae begin feeding on plant roots, burrowing below the frost line and spending winter there. The life cycle takes a year but can extend to two years in the northern parts of the range.

Exclusion and Prevention

➤ Reduce lawn watering in late summer when females are laying eggs to desiccate the eggs and hatching larvae.

➤ Use floating row covers.

- Install pheromone traps.

   Note: Studies have shown that pheromone traps actually attract more beetles than they trap. If you use traps, locate them well away from valued plants and use a secondary control to eliminate them.

Japanese Beetles and leaf damage

- Knock adults from plants into a cloth, vacuum or handpick and either squash or drown in soapy water. Beetles are easier to pick off in the cool early morning because they are more lethargic.

- Remove sickly garden trees and plants because diseased plants are targets for beetles.

- Promote community-wide trapping. Beetle trapping is most effective if done on a community basis with many traps spread over a large area.

- Use landscape and garden plants that are resistant to Japanese beetles such as begonia ageratum, coreopsis, dusty miller, foxglove, hosta, impatiens, lantana, nasturtium, pansy, poppy, and violet.

CONTROLS

Insecticidal soap	Pyrethrum
Neem oil	Spinosad

BENEFICIAL BIOLOGICALS

Bacillus thuringiensis (grubs)

Beauvaria bassiana

Milky spore *(Bacillus papillae)* (grubs)

Nematodes, beneficial (*Heterorhabditis*

bacteriophora, Steinernema carpocapsae, or *S. glaseri* [grubs])

Wasps, parasitoid (*Tiphia vernalis,* naturalized in the United States but not available commercially)

Interesting Facts

Many states have quarantines on the movement of plants to help stop the spread of Japanese beetles.

   Other beetles produce grubs that appear similar to Japanese beetle grubs such as June bugs or May beetles.

Beetles:
Mexican Bean Beetles
(Epilachna varivestis)

How common are these pests?

The Mexican bean beetle is found in most states east of the Mississippi River and also in Arizona, Colorado, Nebraska Texas, and Utah, but it is most damaging and most abundant in the southernmost part of its range.

What do they look like?

Adult beetles are oval and domed and about ¼ inch long. Coloration ranges from a pale yellow to a reddish orange with 16 black spots arranged in three rows of 6-6-4 across the wing covers. The Mexican bean beetle lacks the white area between the head and the body that is present in its relative, the lady beetle. Larvae are fat yellow grubs, about 5/16 inch long, with long, branching, dark spines distributed over their body. Eggs are yellow and laid on the undersides of leaves in groups, placed upright on end.

Adult Mexican Bean Beetle

What kinds of plants do they attack?

The most common hosts are beans, such as cowpeas, lima beans, pole and snap beans, soybeans, and string beans, but Mexican beetles will sometimes feed on cabbage, kale, and mustard greens.

Where are they found on the plant?

Adults and larvae stay on the undersides of leaves, stems, and seedpods of beans.

What do they do to the plant?

Feeding results in skeletonized areas and sometimes whole leaves left with nothing but the veins intact. The leaves turn brown and die.

General Discussion

Adults overwinter in field or garden debris or in nearby fields or woods. In late spring, the female emerges over an extended period and lays eggs on the lower leaf surfaces, as many as 450 in her lifetime. The eggs hatch in five to 14 days, and the larvae will feed for two to four weeks before pupating on the leaves in small groups. Adults emerge in about 10 days. One to three generations are produced annually. Higher numbers are produced in the southernmost part of their range. The greatest injury to plants usually occurs in late summer when the population reaches its peak.

Exclusion and Prevention

Mexican Bean Beetle Larvae

➤ Vacuum or handpick adults and larvae from plants and squash or shake them off into a tarp for disposal.

➤ Plant resistant varieties such as the snap bean varieties Black Valentine, Logan, and Wade, which are less severely damaged by the beetles.

➤ Plant fast-maturing, early-season varieties to avoid the main surge of adults in late summer.

➤ Use floating row covers.

➤ Plant soybeans as trap crops for Mexican bean beetles, then destroy the plants when they are heavily infested.

➤ Attract natural predators and parasites by leaving flowering weeds in between the garden rows or by interplanting flowering plants and herbs.

➤ Remove garden debris at the end of the season to eliminate overwintering sites.

CONTROLS

Citrus oils

Herbal oils

Insecticidal soap

Neem oil

Pyrethrum

BENEFICIAL BIOLOGICALS

Assassin bugs

Beauvaria bassiana

Bacillus thuringiensis var. *san diego*

Lady beetles

Minute pirate bugs

Nematodes, beneficial
  (*Heterorhabditis bacteriophora*)

Spined soldier bugs

Wasps, parasitoid *(Pediobius foveolatus)*

Interesting Facts

A word of caution about squashing Mexican bean beetle adults and larvae: the adults exude a yellow, foul-smelling liquid from underneath their body that discolors skin. This is a defense mechanism against birds or other insects to make themselves an unappetizing meal. While foul smelling, the liquid is not harmful. The spines that cover the larvae are not harmful and not sharp, despite their menacing appearance. This, too, is an effort to deter predators.

Mexican bean beetles are relatives of the beneficial lady beetles. It is one of the few members of this family that feed on plants rather than other insects.

Caterpillars

Cabbageworms

Corn Borers

Cutworms

Squash Vine Borers

Tobacco Budworms

Tomato Hornworms

Caterpillars are the larvae of moths and butterflies. Different species vary widely in their diets, forms, habits, and length of time spent as larvae. Some are host specific while others are generalists.

Only a few millimeters long when they hatch, caterpillars are the fastest-growing larvae of the insect kingdom. By the time they prepare to pupate most increase their size a thousand-fold. Some grow ten-thousand-fold from their original size.

Caterpillars have two goals: to not get eaten and to grow fast so they can enter the next stage of life as pupas. This requires constant eating. They grab plant tissue in their mouths, tear or bite it off, and chew. Even a single caterpillar can damage a seedling or in cases of borers, bring down a large plant. Eventually they've eaten enough. Each caterpillar spins a cocoon and undergoes metamorphosis, transforming into a completely new structure, emerging from the cocoon as a butterfly or moth.

Tomato budworm climbing a stem

Caterpillars are eating machines. Their tubular bodies consist mostly of intestines that digest quickly and efficiently. Their mandibles (jaws) are adapted to reduce plants into meals. Depending on species they may feed on leaves or other living plant parts, decaying leaves, fungus, pollen, or spores. They may be daytime feeders or nocturnal. For instance, cutworms hide near plants during the day and feed at night, often surprising gardeners with a tray of savaged seedlings.

Bacillus thuringiensis (Bt-k) targets caterpillar, moth, and butterfly larvae. It is very safe to use and is the active ingredient in Garden Dust and Caterpillar Killer from Safer Brands.

Caterpillars:
Cabbageworms

How common are these pests?

The adult cabbageworm butterfly is the most common butterfly in most gardens in North America. It is found everywhere.

Cabbageworm larva

What do they look like?

Adults are butterflies with a 1½- to 2-inch wingspan. Their forewings are white with black tips. The wings of males have a single black spot while those of females have two. Larvae are deep velvet green, sluggish caterpillars. They have five pairs of prolegs. Older larvae have a yellow line that runs the length of their back. Eggs are yellow and cone shaped and are laid on the undersides of leaves.

What kinds of plants do they attack?

All members of the cabbage family are hosts, including broccoli, Brussels sprouts, cabbage, cauliflower, collard, horseradish, kale, kohlrabi, mustard, nasturtium, radish, sweet alyssum, turnip, and many related weeds.

Where are they found on the plant?

The adults can be seen flitting around the garden occasionally stopping to drink nectar from a flower or to lay eggs on the undersides of leaves. The caterpillars stay on the foliage.

What do they do to the plant?

The adults do not feed on plants but instead drink their nectar. The larvae feed on foliage. Larvae can reduce whole plants to bare stems and veins in short order. They prefer leafy

Protect Your Garden

foliage but burrow into the heads of cabbage and Brussels sprouts as they mature. Larvae also leave behind large amounts of excrement that makes harvested produce undesirable.

General Discussion

Female cabbageworm moth. Males have only one spot on their wings.

Cabbageworms are members of the Pieridae family, which contains other similar moths such as the white and sulfur moths. The imported cabbageworm was first found in Quebec City, Canada, in 1860 and quickly spread throughout most of North America. Until cabbageworms were introduced, other species of cabbage moths dominated fields and gardens, including the alfalfa cabbageworm and the southern cabbageworm, but now *Pieris rapae* dominates.

Cabbageworms overwinter as pupae in the previous year's plant debris, and adults emerge in early spring to lay yellow, bullet-shaped eggs on the lower surfaces of outer leaves. Females lay from 300 to 500 eggs in their lifetime. The eggs hatch in three to five days. The larvae prefer the outer leaves at first and then, as they mature, move to the heads of plants. In two to three weeks the larvae are full grown and drop to the soil to pupate in the debris there or pupate on the host plant. Each larva pupates in a chrysalis that varies in color from yellow to brown or gray. In areas where multiple generations occur, adults emerge in about 11 days. Adults may live up to three weeks. They are most active in the daytime.

Exclusion and Prevention

► Vacuum or handpick larvae and squash eggs.

► Use floating row covers—cabbage crops do not need pollination so the covers need not be removed for pollination.

► Sprinkle crop leaves with cornmeal or rye flour. When the larvae eat the flour, it expands in their gut as it absorbs water and they die.

► Turn over garden soil at the end of the season to expose any overwintering chrysalises to predators and winter freezes.

► Spray plants with garlic or hot-pepper spray weekly as soon as butterflies appear.

► Control weeds around the garden area, especially weeds in the mustard family.

- Plant resistant varieties of cabbage such as Early Globe, Red Acre, and Round Dutch.

- Soak harvested produce in warm salt water before cooking. Any larvae inside will die and float to the top of the water.

Cabbageworm damage on cabbage leaf

CONTROLS

Black pepper spray

Citrus oils

Herbal oils

Neem oil

Pyrethrum

Spinosad

Wood ashes

BENEFICIAL BIOLOGICALS

Assassin bugs

Beauvaria bassiana

Bacillius thuringiensis var. *kurtaski*

Damsel bugs

Green lacewings

Hoverflies

Nematodes, predatory
  (*Steinernema carpocapsae*)

Tachinid flies

Wasps, parasitoid (*Trichogramma* wasps)

Cabbageworm butterfly sipping nectar from cabbage flowers. Only the larvae harm the plant.

Interesting Facts

These butterflies regularly appeared in early gardens about the time cows begin to produce milk in spring. That's the source for the English term "butterfly."

Several crucifer pests are similar to the cabbageworm, among them the cabbage looper. The adult of this species *(Trichoplusia ni)* is a night-flying moth. The larvae arch their backs and crawl along inchworm-style. They, too, can cause severe damage by eating holes in leaves and defoliating whole plants.

Caterpillars:
Corn Borers (European)
(Ostrinia nubilalis)

How common are these pests?

European corn borers are common throughout the United States and southern Canada east of Rockies with the exception of Texas and the southern portion of Florida.

What do they look like?

The female mates several times, then lays an average of 30 creamy white eggs in clutches ¼ inch in diameter on the undersides of leaves. She lays eggs twice a night for 10 days. The eggs hatch in three to seven days depending on temperature. Individual eggs are smaller than 4/100 inch (1 mm). The larvae are tan to light gray, with slightly darker stripes running down their slender length. The heads are dark brown. The pupae are reddish brown, torpedo shaped, and about an inch long. Time from egg lay to sexual maturity is about 30 to 45 days depending on temperature. Mature female

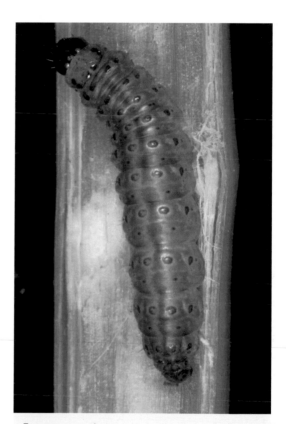

European corn borer

moths are an inch long with a 1-inch wingspan. They are whitish to yellow brown with irregular brown wavy bands stretching across the wings. Males are a little smaller and a bit darker.

What kinds of plants do they attack?

Larvae prefer corn and other grains such as hops, sorghum, sunflower, and other plants

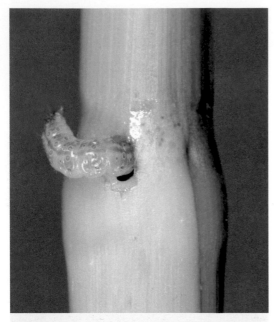

Larva of the European corn borer

with long stems that they can bore into. However, they can be generalist, attacking asters, beans, celery, chrysanthemums, dahlias, gladiolas, peppers, potatoes, Swiss chard, and many other species.

Where are they found on the plant?

The larvae prefer to embed themselves inside the stem if possible, but they can be found on the outside of some plants. They attack leaves, stem, and ears of corn.

What do they do to the plant?

Plant leaves are chewed to the point that they fall from the plants. Once inside the stem, the larva disrupts transport of liquids. The area above the invasion quickly withers.

General Discussion

European corn borers are often far more damaging in their southern range, in the southern tier of the United States, where the borers are not thwarted by cold weather and can complete four crops a year. In the northern range, 50 miles north into Canada to 100 miles south, they complete only one generation per year so the population doesn't reach the same density.

Exclusion and Prevention

Eggs overwinter in crop residues so removing, covering, or composting them reduces initial infestation rates. Remove weeds, especially artemisia.

➤ Use row covers.

➤ Handpick larvae.

➤ Install light traps to capture reproductive adults.

European corn borer eggs

CONTROLS

Black pepper spray

Citrus oils

Herbal oils

Neem oil

Pyrethrum

Spinosad

Wood ashes

BENEFICIAL BIOLOGICALS

Assassin bugs

Beauvaria bassiana

Bacillus thuringiensis var. *kurtaski*

Damsel bugs

Green lacewings

Hoverflies

Nematodes, predatory
  (*Steinernema carpocapsae*)

Tachinid flies

Wasps, parasitoid (*Trichogamma* wasps)

Adult southwestern corn borer moth

Interesting Facts

Before corn was introduced to Europe, the corn borer's preferred crops were grains, hemp, and hops.

Caterpillars:
Cutworms

How common are these pests?

The term "cutworms" refers to the larvae of several moth species belonging to the family Noctuidae. They are common in most gardens.

What do they look like?

Adults are gray moths with mottled patterns on their wings and a wingspan of about 1½ inches. Larvae are plump gray, green, brown, or striped caterpillars that curl into a characteristic C-shape when disturbed. They are 1 to 2 inches long to maturity and have a shiny head.

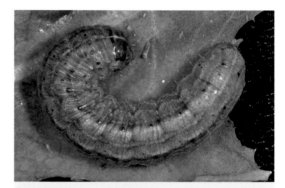

Larva of the army cutworm

What kinds of plants do they attack?

Cutworms attack seedlings, but some cutworms climb plants to feed on buds, fruit, and leaves. Others remain underground and feed on roots. Some of their favorite hosts are beans, cole crops, corn, cosmos, hops, snapdragons, sunflowers, and tomatoes.

Larva of variegated cutworm

Where are they found on the plant?

Most cutworms live in or on the soil during the day and come out at night to feed. Adults do no damage to plants and only fly at night, sipping nectar from plants.

What do they do to the plant?

The cutworms most commonly found in gardens sever young seedlings at the base by girdling the stem or chewing all the way through, toppling the plant. Cutworms ruin more plants than they consume as once they have felled a seedling, they move on to the next victim, leaving behind a path of severed plant tops.

General Discussion

Females mate and lay eggs on grass or plant surfaces. If they are laid early enough, larvae emerge and overwinter in plant debris or in the soil as partially mature larvae, so they are robust and hungry when seedlings are transplanted in spring. Eggs laid later overwinter. As temperatures warm, the cutworms become active and hunt for food. Once they are fully grown, which takes from two to six weeks depending on species, temperature, food abundance, and quality, they pupate in the soil for several weeks and emerge as adult moths. The adults climb out of the pupation chamber using tunnels they created as larvae. If the tunnel is blocked on the surface, the moth cannot emerge and dies.

Eggs of the western bean cutworm

After emerging, the moths mate and begin laying eggs.

Depending on the environment, cutworms produce one to several generations per year.

Exclusion and Prevention

➤ Remove early weeds that serve as food sources for the emerging caterpillars.

➤ Cultivate the garden in the early spring to expose and unearth the larvae and delay planting to starve them.

➤ Protect seedlings. Start seeds in pasteurized soil in containers or trays out of harm's way, such as on a table or placed in moats.

➤ Cover seedlings overnight with tray covers.

➤ Place paper cups, tin foil, or paper collars around new transplants to block cutworms' access. Sink the cups or collars an inch into the ground and extend several inches up the stems. Toothpicks placed on either side of the stems also deter cutworms.

➤ Place a cloth or porous but impervious plastic sheet over the soil to stop the cutworms from emerging from the soil in spring or dropping into the earth to pupate later in the season.

- Plant sunflowers around the perimeter of the garden as a trap crop, then destroy the attacking cutworms daily.

- Use floating row covers.

- Place bat houses near the garden. Bats are fond of the adult moths.

- Turn the garden soil after harvest to expose eggs, cutworms, and pupae to bird, insect, and rodent predators or to bury them deeply.

Army cutworm, adult

CONTROLS

Alcohol spray

Black pepper spray

Herbal oils

Insecticidal soap

Neem oil

Pyrethrum

Spinosad

Wood ashes

Set out cornmeal in shallow saucers throughout the garden. Cutworms eat the cornmeal, bloat, and die because they cannot digest it.

BENEFICIAL BIOLOGICALS

Assassin bugs

Bacillus thuringiensis var. *kurstaki*

Damsel bugs
(*Nabis sp., a* native predator)

Ground beetles

Nematodes, parasitic
(*Steinernema carpocapsae* or *S. feltiae)*

Wasps, parasitoid *(Trichogramma* wasps)

Interesting Facts

Several species of cutworms that feed aboveground are known as "armyworms." These cutworms often occur in great numbers and move in large migrations, hence the name.

Caterpillars:
Squash Vine Borers
(Melittia cucurbitae)

How common are these pests?
Squash vine borers are found throughout the United States, east of the Rockies.

What do they look like?
Adult squash vine borers are clear-winged moths with red and black abdomens 1 to 1½ inches long. The back legs are fringed with feathery black and orange hairs. The larvae are white with brown heads and are 1 inch long. Eggs are small and dull red and are glued to the base of the plant and on the undersides of leaves.

What kinds of plants do they attack?
Summer and winter squash are favored plants, but squash vine borers will also attack other cucurbits such as cucumbers, gourds, melons, and pumpkins.

Where are they found on the plant?

Squash Vine Borer

Larvae do the damage and are found in the tunnels they create at the base of the vines and stems of their hosts.

What do they do to the plant?

Squash vine borers usually make tunnels, which cut through the inner core of the stem, disrupting growth and causing wilting and the plant's eventual death. You can find the pinhole where the tunnel entered the stem and follow its journey through the plant.

Larvae also girdle vines and stems, which also causes wilting and death. Yellow frass is left behind in the tunneled stems. Pathogens colonize the frass, further weakening the crippled plant. Often the first sign of an infestation is wilting plants. By this time it is often too late. Wilting often occurs only under full sun at first, but gradually the plants wilt and do not rebound.

General Discussion

Squash vine borers spend winter underground in cocoons as fully mature larvae or pupa. Adults emerge in mid-spring and mate, and the females lay eggs at the base of host vines and stems. Eggs hatch in about seven days, and the larvae bore into the plant to feed. Larvae feed for six weeks; then they emerge from the plant, burrow into the soil, and pupate. In the southern part of their range, two generations are possible.

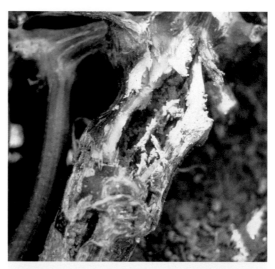

Squash vine damaged by squash vine borer

Exclusion and Prevention

➡ Plant early to avoid the surge of emerging adults at mid-spring.

➡ Use floating row covers.

➡ Handpick larvae and squash eggs.

➡ Slice open affected vines and kill the borers.

➡ Mound soil around the base of plants up to the first flower to exclude females searching for places to lay their eggs and to promote adventitious roots.

➡ Plant resistant varieties of butternut squash, cucumbers, melons, summer crookneck squash, and watermelons.

- Cover the soil with a barrier to stop the pupae from dropping into the soil and to keep females from laying eggs.

- Wrap the stems with sticky tape to catch pregnant moths that wish to lay on the stems.

- Wrap the stem with aluminum foil or paper as barriers.

CONTROLS

Black pepper spray	Neem oil
Herbal oils	Pyrethrum
Horticultural oil	Spinosad
Insecticidal soap sprayed weekly on the base of the plant	Wood ashes

BENEFICIAL BIOLOGICALS

Beauveria bassiana

Btk *(Bacillus thuringiensis* var. *kurstaki)* injected into the stem near the borer hole.

Nematodes, beneficial: Both *Heterorhabditis bacteriophora* and *Steinernema feltiae* attack pupae in the soil. They can also be injected into the stem to attack the burrowing caterpillar.

Interesting Facts

The key to controlling this insect is to affect the larvae before they enter the stems. Once inside, they are safely ensconced and treatment is difficult.

Squash vine borer at base of damaged squash plant. If damaged plants are split open, the white, 'grub-like' caterpillar can be found.

Caterpillars:
Tobacco Budworms
(Heliothis virescens)

How common are these pests?

Tobacco budworms, or geranium budworms as they are sometimes called, are found primarily in the southern tier of the United States because neither adults nor pupae can survive cold winters. However, in years with mild winters, adult moths migrate to northern gardens where gardeners may notice damage in late summer or early fall.

What do they look like?

Adults are light brown moths with wingspans of about 1½ inches with dark stripes on their upper pair of wings. Larvae are 1 inch long and may vary in color from pale green to black, brown, red, or even black. Some individuals have stripes and others do not. Their color variation is thought to be due to diet and life stage. The presence of microspines on some segments of the larval body is used to differentiate budworms from their close relative the corn earworm.

Tobacco budworm

What kinds of plants do they attack?

Larvae damage the buds of many different crops and flowers, including ageratum, alfalfa, chrysanthemum, cotton, geranium, marigold, nicotiana, petunia, snapdragon, soybean, tobacco, verbena, zinnia, and many common garden vegetables.

Where are they found on the plant?

Larvae bore into the flower buds of plants and may eat leafy tips if flower buds are not present. Adults do not harm plants.

Tobacco budworm moth

What do they do to the plant?

Small larvae tunnel into flower buds, leaving behind characteristic holes in the bud. Older larvae consume the entire bud and chew the petals of opened flowers. The foliage is usually left unharmed, but the chewed or damaged flower buds reduce production in crop plants and ruin the appearance of cut and garden flowers.

General Discussion

Tobacco budworm pupae overwinter a few inches beneath the soil, snugly ensconced in an earthen chamber. During early to late spring, adult moths emerge from the ground and seek out mates, becoming most active in the early evening. Females lay eggs in clusters of 18 to 25 on flower buds and leaves. The eggs are spherical with a flat end and narrow ridges. Newly laid eggs are pale yellow to white and darken as they approach hatching. Young caterpillars emerge and begin to feed on the bud tissue for about three weeks. Now fully grown, they drop to the ground to pupate.

Exclusion and Prevention

➤ Plant ivy geraniums instead of common geraniums.

➤ Handpick caterpillars from plants and out of flower buds and destroy. Do the picking at dusk when the larvae are most active.

➤ Use a bug zapper hung over the favored host plants to attract and kill the adults. Be aware that bug zappers will also kill beneficial insects.

➤ Hang wire, cone-shaped traps baited with pheromones specific to budworms a short distance away from the host plants.

➤ Use row covers.

➤ Lay ground barriers such as rugs or agricultural cloth to prevent the pupae from emerging.

CONTROLS

Black pepper spray	Pyrethrum
Herbal oils on the caterpillars	Spinosad
Neem oil	Wood ashes

BENEFICIAL BIOLOGICALS

Assassin bugs	Minute pirate bugs
Big-eyed bugs	Nematodes, predatory
BtK	*(Steinernema carpocapsae)*
Damsel bugs	Wasps, parasitoid

Interesting Facts

Tobacco budworms can devastate agricultural crops but are more of an annoyance for home gardeners. In the cut-flower industry they can devastate a crop by damaging existing blooms and by reducing the plant's ability to produce new flowers. Milder winters increase the likelihood that tobacco budworms will become an increasing problem in northern areas.

Caterpillars:
Tomato Hornworms
(Manduca quinquemaculata)

How common are these pests?

These goliaths of the garden are found throughout the United States.

What do they look like?

Adults are large moths, usually gray or brown, with wingspans of 4 to 5 inches. Their large forewings are mottled with swirls of light and dark, and their heavy bodies are marked with two lines of yellow spots. The larvae are the real stars, reaching lengths of up to 4½ inches. Their bodies are thick, about ½ to ¾ inch in diameter, and are vivid green with eight white, chevron-shaped marks down their sides and several dark spots that look like eyes. Their most prominent feature is the black horn that protrudes from the last segment of their rear.

What kinds of plants do they attack?

Tomato hornworms feed only on plants in the nightshade family, including its namesake tomato, as well as eggplant, pepper, potato, tobacco, and tomatillo.

Tomato hornworm last stage larva, defecating

Healthy tomato plant

Where are they found on the plant?

Larvae feed on the upper leaves, stems, and fruit of their host plants. Adults do not harm the plants and feed only on plant nectar.

What do they do to the plant?

Larvae are voracious eaters. They can defoliate a plant in a few days, consuming whole leaves and leaving behind dark green ¼-inch-long excrement pellets that accumulate in massive amounts. This excrement is often seen before the worms are seen. Green fruit is preferred over ripe, and young plants may be completely consumed.

General Discussion

Tomato hornworms overwinter as pupae in an earthen chamber 4 to 6 inches deep in the soil. In mid- to late spring, adult moths emerge and fly long distances in search of mates and food. Eggs are round and white and are laid singly on the top and undersides of the host. Eggs hatch in about a week. The larvae pass through four or five molts in the month they feed. The first instar or molt (each stage is numbered by the shedding of its old skin) is yellow to white and has no markings. Later instars turn green and develop the characteristic white chevron stripes. When grown, they drop to the soil and burrow down to create the cell for pupation. One or two generations may be produced in a year.

Exclusion and Prevention

➤ Handpick and destroy the caterpillars. Hornworms tend to feed near the outside foliage of plants in the late afternoon and at dawn, making them easier to find despite their effective camouflage.

➤ Till the garden soil after harvest to kill buried pupae or expose them to birds, rodents, and other predators.

CONTROLS

Black pepper spray	Pyrethrum
Herbal oils	Spinosad
Insecticidal soap	Wood ashes
Neem oil	

Do not destroy worms found with white rice-like sacs hanging off their bodies. These are the pupae of a naturally occurring parasite, the Brachonid wasp *(Cotesia congregates)*. The larvae of these wasps, hatched from eggs laid on the worm, will feed on the hornworm until they are ready to pupate. Adult wasps that emerge from these white sacs will lay eggs on other hornworms and establish a parasite population in the garden.

See the wasps in action: youtube.com/watch?feature=player_embedded&v=sjXf_kCZp50

A wasp has injected eggs into this hornworm and they have emerged through the body. When the eggs hatch into larvae, the caterpillar will be eaten.

BENEFICIAL BIOLOGICALS

Assassin bugs

Btk

Green lacewings

Lady beetles

Nematodes, predatory
(Steinernema carpocapsae)

Wasps, parasitoid

Interesting Facts

Tomato hornworms belong to a family of moths known as hawk, hummingbird, or sphinx moths. The tomato hornworm moth flits and dips among plants, sucking up nectar with its proboscis and fooling many gardeners into thinking it is a hummingbird instead of a ravenous hornworm.

The tobacco hornworm *(Manduca sexta)* is also a pest of tomatoes and, depending on whom you ask, can cause as much damage as the tomato hornworm. Tomato hornworms are uncommon along the Gulf

Tomato hornworm moth

Coast, but tobacco hornworms are very common in that area. These close relatives to the tomato hornworm are also large and green but have seven white diagonal lines that are more of a slash than the chevron shape found on tomato hornworms. Tobacco hornworms have a red horn on their rear.

Deer

How common are these pests?

Deer populations vary widely, both geographically and by habitat. Generally, they prefer light forest and grasslands near forested areas to provide them a quick retreat if threatened while grazing in the open. Deer often visit suburban gardens near forested areas.

What do they look like?

Deer are grazing mammals with graceful bodies, thin legs, and long necks with small heads and narrow muzzles. They vary in size depending on species, sex, age, and locale, but they are usually between 4 and 6 feet long and weigh between 80 and 220 pounds. Deer grow larger in the north than in the south. Males grow antlers beginning in late summer, shedding them in late winter through spring.

Mule deer, named for their large mule-like ears are native to western North America.

What kinds of plants do they attack?

Deer browse on a wide variety of plants, which often changes with the time of year and the availability of natural food. They are particularly fond of both annual and perennial garden and vegetable plants. They are fonder of some species than others, but their tastes widen with scarcity. They prefer young, tender, succulent growth and, given the choice, avoid plant textures that are fuzzy, rough, or strongly scented or flavored.

Where are they found on the plant?

Deer graze on plants from ground level up to about 6 feet. They emerge from the forest at night to browse in yards but will flee quickly when approached. They have excellent hearing and eyesight. Aside from a destroyed garden that is half eaten and knocked down by prancing hooves, they leave piles of small, dark, pelletized droppings.

Black-tailed deer, a subspecies of mule deer, frequent the coniferous forests of the North Pacific Coast

What do they do to the plant?

Deer have no upper front teeth; therefore, they break or tear off leaves and limbs using their lips and lower teeth instead of biting cleanly through them as rabbits or gophers do. This leaves behind ragged browse damage. Deer also trample plants as they feed.

Whitetailed deer

General Discussion

Deer follow a normal mammalian life cycle. Adults mate in late fall through early winter. Fawns are born in late spring to early summer. Twins are rare. Maturity is reached in one to two years. A deer may live for 10 to 20 years if not killed by predators or disease, which is infrequent. Their usual life expectancy past the first year is one to eight years.

Exclusion and Prevention

- Although many scents such as capsaicin, garlic, human hair, scented soap, used clothing, and other repellants may have some success, deer become accustomed to them and after a while will venture past them.

- Only two types of odor repellents have been found to be consistently effective: putrescent eggs and predator scent, such as big cat urine or coyote urine.

- Frightening deer is effective for a short time, but they become accustomed to the signals. High-pressure water sprinklers, bright lights, battery-powered radios, and ultrasonic noises connected to motion sensors are among the keep-out signals. The degree that deer pay attention to the signs depends on food pressure. If dinner is plentiful, deer will leave rather than listen to loud noises when they eat. However, if food is scarce and the meal comes with noise, they will cope with the inconvenience.

➤ Fencing is the most effective barrier. Construct it according to deer habits. They can jump over any fence lower than 8 feet. They also slip under fences unless built tight to the ground.

➤ Electrified fences with two or three strands of 20-gauge wire also deter them.

➤ Cow grating around the perimeter of the garden keeps them from getting close to the plants.

CONTROLS

Purtescent eggs

BENEFICIAL BIOLOGICALS

Some **dogs** and **geese** are effective at keeping the garden clear of deer.

Deer fences should be 8 feet tall and secure enough to withstand deer force.

Electrified fences can be charged with solar power, are easy to install, and are very effective.

Earwigs

How common are these pests?

Introduced to North America from Europe in the early 1900s, earwigs are found throughout the United States except for a few southern states.

What do they look like?

Adult earwigs are reddish brown, glossy, and about ½ inch long, and they have a pair of prominent pinchers extending from the tip of the abdomen. While evil looking, these pinchers are harmless to humans and are used in mating and to manipulate food. The pinchers

European Earwig on *Anemone*

of the male earwig are curved and the female's are straighter. Adults rarely fly, although some species have wings, and they also rarely pinch. Larvae appear similar to the adults.

What kinds of plants do they attack?

Earwigs eat artichokes, bean seedlings, corn silks and kernels, lettuce, potatoes, roses, strawberries, and zinnias. They burrow into bulbs and fruit.

Where are they found on the plant?

Adult earwigs and nymphs are omnivores, preferring meat and decaying organic matter but also feeding on leaves and flowers. They can be found on the ground and climbing on leaves.

What do they do to the plant?

Earwigs rarely cause serious damage to plants and attack plant predators. Their presence can

Earwig, adult

Earwigs in milkweed

be tolerated. They chew ragged holes in the margins of leaves and flower petals that resemble caterpillar damage. Earwig damage is usually minimal to most plants. They can be difficult to remove, hanging on to leafy vegetables despite repeated washings. They feed at night and hide in dark, tight places during the day. Thus they are sometimes harvested with flowers and vegetables.

General Discussion

Earwigs overwinter as adults in garden debris and under stones, boards, or bricks. They become active as temperatures warm in late winter. Females lay about 50 eggs in a small round nest they make beneath a rock or in a protected area. Eggs hatch in April, and the mother earwig protects and nurtures her young for several weeks until they have molted and are ready to leave the nest. They grow into mature adults in about a month. Females sometimes lay a second, smaller clutch of eggs in May or June. Earwigs forage for food at night, making them difficult to spot in daylight.

Exclusion and Prevention

➤ Remove debris or areas for adults to overwinter.

CONTROLS

Diatomaceous earth

Spread Tanglefoot on tree trunks or around plants.

Citrus and other botanical oils

Pyrethrum

Spinosad

Make traps out of items such as damp newspaper, lengths of hose, boxes with small holes cut in the sides, a low-sided tuna cans, or 10-inch-square pieces of cardboard, laid between rows and around plants. Bait the traps with oatmeal, a drop or two of tuna oil or fish oil, or even a drop of meat grease. Earwigs hide in these dark, damp places during the day. Collect the traps and shake out the inmates.

BENEFICIAL BIOLOGICALS

Diatomaceous earth

Spread Tanglefoot on tree
    trunks or around plants.

Citrus and other botanical oils

Pyrethrum

Spinosad

Earwig nymph on flower

Interesting Facts

The name "earwig" is derived from an
old myth that claims earwigs crawl into
the ears of sleeping humans and animals at night, burrowing into the brain. There is no
truth to this.

Earwigs are true omnivores and are also beneficial in the garden as their preferred food
source is insect larvae and eggs, especially aphids. Some species have defensive glands in
their abdomens from which they can squirt a noxious liquid as far as 4 inches if threatened.

Four-Lined Plant Bugs
(Poecilocapus lineatus)

How common are these pests?

Four-lined plant bugs are mainly found in the northeastern quadrant of the United States and adjacent areas of southern Canada, but they also have populations in other parts of the United States.

What do they look like?

Adults are greenish yellow and about ¼ inch long with four black lines running the length of the wing covers. Nymphs are bright red orange with black spots on the thorax and develop the characteristic striping as they reach maturity. These insects move very quickly and can be hard to spot.

What kinds of plants do they attack?

The host range is wide for this insect, including Amur maples, azalea, basil, bluebeard, chrysanthemum, cucumber, currants, forsythia, geranium, gooseberry, lavender, mint, roses, Russian sage, sage, Shasta daisy, squash, sumac, viburnum, and many other flowers and herbaceous plants.

Where are they found on the plant?

On the upper sides of leaves

Four-lined plant bug on cucumber leaf

Four-lined plant bug damage, cucumber leaf

What do they do to the plant?

Four-lined plant bugs are true bugs. They use their straw-like stylets to pierce the leaf surface, then release enzymes that dissolve the tissue for easy juice-sucking.

Feeding sites are very distinctive—perfectly round white or black spots where the leaves are scraped but not pierced through. Leaves eventually turn translucent and die as the cells collapse. Symptoms may not appear until after the insect has left the plant. Damage is usually aesthetic and can be pruned away. Feeding usually never causes the death of plants.

General Discussion

The four-lined plant bug overwinters as small clusters of banana-shaped eggs carefully placed in slits in dormant plant shoots. Nymphs hatch in mid- to late spring and develop into adults by July. Once adults, they spend about a month feeding, mating, and laying eggs before dying. Generally, one generation per year is produced.

Exclusion and Prevention

➤ Since the bug rarely creates extensive damage, it may be tolerated in the garden.

➤ Prune all herbaceous stems to the ground in fall and destroy the prunings.

➤ Use floating row covers.

➤ Vacuum or handpick adults and nymphs.

➤ Plant mint as a trap crop.

CONTROLS

Azadirachtin	Neem oil
Herbal oils	Pyrethrum
Insecticidal soap	Spinosad

BENEFICIAL BIOLOGICALS

Assassin bugs	Damsel bugs
Beauvaria bassiana	Praying mantis
Bacillus thuringiensis var. *kurstaki*	

Interesting Facts

Four-lined plant bug damage can be mistaken for a fungal-foliage disease due to the water-soaked and translucent appearance of the damaged tissue. When the foliage is disturbed, the plant bugs will scurry away, leaving the observer no evidence to connect insects to the damage.

Fungus Gnats

How common are these pests?

Fungus gnats are common in indoor growing areas and are found outdoors in warm, moist areas.

What do they look like?

Fungus gnats are about .13 inches (3- 4 mm) in length, dark grayish to black in color. They are slender insects with delicate long legs and long wings. Larvae are clear to creamy white with a shiny black head and can be up to ¼ inch long.

Fungus gnat (*Leia ventralis*)

What kinds of plants do they attack?

Fungus gnats attack a wide variety of plants. Most are especially annoying pests of houseplants and occasionally lawns. They feed on debris in the planting medium and the roots of indoor houseplants, garden ornamentals, vegetables, and many common plants, including carnations, clover, corn, cucumbers, Easter lilies, geraniums, lettuce, nasturtium, peppers, poinsettias, and potatoes.

Where are they found on the plant?

The larvae of fungus gnats live in the soil, eating algae, organic matter, roots, and root hairs. Adult fungus gnats do not feed and live only to reproduce.

What do they do to the plant?

Fungus gnat larvae damage the epidermis of plant roots and weaken the plant by making roots susceptible to attack. They are also vectors for diseases, including *Botrytis*, *Fusarium*, and *Pythium*.

General Discussion

Fungus gnat adults and larvae live in moist, shady areas, both indoors and out. Adults hover near the surface of the soil, and larvae spend their lives in the soil at the root level. Adult females lay eggs in cracks and crevices on the surface of the growing medium near the plant stem. The larvae hatch and burrow into the soil to feed in the upper 3/8 to 6/8 inch (1 to 2 cm) of soil, eventually penetrating plant roots. They pupate in the soil before emerging and crawling out to fly away and seek mates. Fungus gnats complete their life cycle in two to four weeks.

Fungus gnat larvae in run off water

Exclusion and Prevention

➡ Avoid overwatering. Overly moist soil attracts adults as does soil medium high in peat moss.

➡ Inspect new plants for the presence of fungus gnats before introducing them into the growing environment.

➡ Place barriers, such as cardboard, cloth, paper, container caps, or a layer of sand over the soil. This prevents larvae from burrowing in the growing medium and prevents adults from emerging after pupating.

➡ Let the soil dry between waterings, as moist upper soil surfaces are ideal locations for egg laying. A layer of vermiculite, perlite, or diatomaceous earth helps with this.

CONTROLS

Herbal oils and teas	Neem oil
Horticultural oil	Pyrethrum
Insecticidal soap	Spinosad

BENEFICIAL BIOLOGICALS

Bacillus thuringiensis var. *israelensis*	Mites, predatory (*Hypoaspis* species)
Beauvaria bassiana	Nematodes, predatory
Minute pirate bugs	(*Heterorhabditis* and *Steinernema* species)

Gophers

How common are these pests?

Gophers are a very occasional problem in the garden. They are found mainly in the central and western United States, Florida, and Mexico.

What do they look like?

Gophers are medium-size rodents ranging from about 5 to nearly 14 inches long, not including the tail. Their bodies are covered with fine fur that ranges from pale brown to nearly black. Their head is small and flattened with small ears and eyes and very prominent incisor teeth. Their forepaws have strong claws that they use to dig.

Gopher

What kinds of plants do they attack?

Gophers eat a variety of roots of flowers, grasses, and vegetables, but they are particularly fond of alfalfa, bamboo, bulbs, bushes, dandelions, and roses.

Where are they found?

Gophers tunnel under gardens and lawns.

What do they do to the plant?

Gophers feed on plants in three ways: first, they eat the plant roots they encounter as they dig tunnels; second, they feed on nearby vegetation they find by venturing short distances from their tunnel entrances; or third, they pull plants into their tunnels from underground. Gophers attract badgers and moles, which consider them prey. The predators cause damage by digging for the gophers.

General Discussion

The presence of gophers in the garden is indicated by mounds of soil that they pile up at the entrances to their tunnels. However, gophers aren't the only mound-making pests in the garden. They are often confused with moles. Mole hills are rough cones with a hole or earthen plug near the center. Gopher

Gopher mounds

mounds are fan shaped with the hole or plug near an edge.

Moles are rarely serious pests in the garden and are not worth the eradication efforts—besides, they eat gophers. They may nibble on a few plant roots but usually aren't detrimental pests. Gophers are more serious pests that chew plant roots, causing wilting and plant death. They usually chew off the entire root system. If plants can be pulled up with only a small tug or if whole plants are gone, the culprit is probably a gopher.

Gophers mate once a year in spring and produce litters of up to five in late spring or early summer. Adult gophers live up to 12 years but average a shorter life because of predators and other environmental factors.

Exclusion and Prevention

➤ Containers, indoor growing areas, and protected raised beds are immune from gophers. In gardens or fields, minimize weeds as they attract gophers.

➤ Line the bottom and sides of planting holes and the bottom of raised beds with hardware cloth (thin metal mesh).

➤ A row of oleander plants around the garden repels gophers. Oleander is a very poisonous plant.

➤ Commercial gopher repellents are available. Most include capsaicin, castor oil, or garlic. Placing these repellents in the mouth of the tunnel may drive them off.

CONTROLS

➤ Repellents are the best way to deal with gophers, but they can be stubborn pests and sometimes the only control is eradication.

- Fumigation is the simplest means of eliminating gophers. Commercial fumigants are generally paper or cardboard cartridges filled with charcoal and potassium nitrate. They are ignited and dropped into the tunnel openings where they produce gases that kill the gophers. Wisps of smoke rising from the surrounding area may signal other exits from the tunnel system. Seal these exits or the gopher will flee.

- Carbon dioxide is another fumigant. Place the CO_2 tank's delivery hose directly in the tunnel opening and turn it on to deliver. Dry ice can also be used. Drop 8 to 16 ounces of dry ice into the tunnel.

- Trapping is effective if fumigation is not. Traps and instructions for them are available at garden shops.

BENEFICIAL BIOLOGICALS

None

Grasshoppers

How common are these pests?

Grasshoppers are found all over North America, where more than 11,000 species have been identified. Most of the time damage from grasshoppers is minimal. However, when their population balloons they become a force of nature to be reckoned with, denuding gardens and fields.

Grasshopper

What do they look like?

Grasshoppers range in size from ½ inch to over 4 inches. Their color varies as an adaption to the environment: from green in grasslands to tan and brown in drier areas. Grasshoppers have a distinctive streamlined body with two armored wings not used for flying and a pair of powerful wings for flying. Their hind legs are capable of propelling them 20 times their body length. The mouth features a mandible, a movable lower jaw with which the grasshopper tears and chops the food, for instance, a leaf from the plant. The grasshopper's lips guide the food into its mouth, which slightly macerates the food before moving it through the digestive system.

What kinds of plants do they attack?

Grasshoppers prefer grasses and grains but will eat any vegetation available when there is food pressure.

Where are they found on the plant?

They can be found anywhere on a plant but prefer middle and upper leaves of taller plants.

Grasshopper (*Melanoplus viridipes*)

Grasshopper – *Melanoplus differentialis* (nymph)

Grasshopper

Mating pair of two-striped grasshopper

What do they do to the plant?

They are capable of eating any soft plant tissue so they can strip a plant when the population is dense. When there are only a few plants, they tear pieces off leaves and eat young shoots.

General Discussion

Grasshoppers are usually not a problem in the garden because they are not especially interested in fruits and vegetables. It is only when the grasshopper population starts increasing, resulting in food pressure, that they become a threat.

Exclusion and Prevention

➤ Use row covers to exclude grasshoppers. If you fear that they may chomp through cloth, use a stronger screen material.

➤ Capsaicin and cilantro repel grasshoppers. Cilantro and hot pepper containing capsaicin can be planted in the garden. Teas made from these spices can be sprayed on plants. Use a sprayer that produces a fine mist to repel them.

➤ Clear land close to the garden. It creates a natural barrier so they land on bare earth or gravel when they jump.

➤ Till the top 2 inches of soil after harvest, exposing the eggs in hardened cases that the grasshoppers laid.

CONTROLS

Neem oil

Pyrethrum

BENEFICIAL BIOLOGICALS

Beauveria bassiana

Domesticated fowl such as chickens, ducks, and turkeys

Nematodes, beneficial

Nosema locustae, a protozoa that infects and kills grasshoppers and locusts is available in commercial products.

Interesting Facts

When some grasshopper populations reach a critical level, their body form changes to a migrating, gregarious form and the grasshoppers move as a wave or storm cloud, denuding all vegetation they encounter. The last time this occurred in the United States was the 1873–1877 swarm of Rocky Mountain locusts in the plains states. Within 30 years the insect became extinct after agricultural tilling destroyed its home breeding areas. Swarms still exist in Africa and the Middle East.

Lace Bugs

How common are these pests?

Lace bugs are found throughout North America, attacking many plants. There are many species. Each feeds on a small family of plants, mostly trees, and perennial bushes. Plants attacked include alder, ash, avocado, azalea, birch, *Ceanothus*, coyote bush, fruit trees, poplar, sycamore, toyon, walnut, and willow. The most damaging species occur in the eastern United States. Most of the time the bug damage is slight and

Sycamore lace bug larvae

should be tolerated because naturally occurring predators and parasites keep damage to a minimum.

What do they look like?

Most lace bugs are about ⅛ inch long with iridescent wings that they hold flat over their back. The intricate venation of these delicate wings gives rise to the name lace bug. Some species also have a lace-like collar over their heads. Most species have some color on their bodies that is species specific. Larvae are more oval shaped than their parents, with spikes on their backs. Adults and nymphs occur together in groups on the undersides of leaves.

What kinds of plants do they attack?

The host range of the many species of lace bugs is wide and varied, but they feed primarily on landscape plants such as azaleas, birch, oak, rhododendron, sycamore, and walnut. However, there are species that feed on chrysanthemum, eggplant, goldenrod, and lantana. Each species feeds exclusively on one type of plant or species closely related to it.

Where are they found on the plant?

Lace bugs feed primarily on the undersides of leaves.

What do they do to the plant?

Lace bugs scrape the outer cells of the undersides of leaves, then suck the juices that leak out. This results in white or yellow stippling on the upper leaf surface. They also produce an abundance of varnish-like excrement around the feeding site called tar spots. A heavy infestation can result in early leaf drop for deciduous plants.

Sycamore lace bug

General Discussion

Lace bugs belong to the insect family Tingidae. They are generally divided into two groups: those that feed on deciduous trees, such as the oak lace bug and the birch lace bug, and those that feed on evergreens, such as the azalea lace bug. Depending on the species, they overwinter either as adults under tree bark and fallen leaves (the deciduous tree group) or as eggs inserted into needle or leaf tissue or glued in place and covered over with excrement (the evergreen group). Eggs hatch into nymphs in late spring and molt several times before becoming adults in about six weeks. Several generations may be produced each year, depending on the species and environment.

Exclusion and Prevention

➡ Plant resistant varieties.

➡ Avoid planting shade-loving plants in full sun, as lace bugs prefer a sunny location and stressed plants.

➡ Knock adults and nymphs off plants early in the season with a forceful stream of water. Nymphs will die before climbing back up to feeding sites.

CONTROLS

Azadirachtin	Horticultural oils
Herbal oils	Insecticidal soap

| Neem oil | Spinosad |
| Pyrethrum | |

BENEFICIAL BIOLOGICALS

Assassin bugs	Lacewings
Beauvaria bassiana	Ladybugs
Bt-i	Minute pirate bugs

Interesting Facts

Of all the species of lace bugs, the azalea lace bug *(Stephanitis pyrioides)* causes the most damage in landscape plants. They can appear in great numbers very quickly and cause widespread bleaching and stippling of the leaves. Once foliage is damaged by lace bugs, it will not recover.

Mite or thrips symptoms can often be mistaken for lace bug damage on the upper leaf surface. However, turn over the leaves and the dots of black excrement distinguish lace bug damage from all other competitors.

Leafhoppers

How common are these pests?

Leafhoppers are common pests in gardens and greenhouses throughout North America. There are many species, and most of them have a varied range of hosts. They may not appear for several years and then reappear in great numbers. They are members of the order Hemiptera, true bugs, and can surge in numbers in a matter of weeks.

Striped leafhopper

What do they look like?

Superficially, leafhoppers look a lot like grasshoppers. They range in length from about ⅛ to ½ inch. They have two pairs of wings, both of which are folded along their back. Two large compound eyes are placed forward near the stylets in the front of the body. When disturbed, they crouch down, which compresses their powerful hind legs that propel them in a fast jump once their energy is released.

What kinds of plants do they attack?

Although different leafhopper species have different food preferences, they are not picky suckers. As food pressure increases they widen their diet.

Where are they found on the plant?

Leafhoppers congregate on the tops of leaves.

When disturbed, they hop off the plant in a flurry that cannot be missed.

Leafhopper (*Acanalonia conica*)

Leafhopper — Sharpshooter *Draeculacephala sp.*

Leafhopper — Striped *Tylozygus bifidus*

What do they do to the plant?

Leafhoppers have piercing-sucking mouthparts, collectively called rostums. The first pair has sharp teeth that are used to scrape the leaf surface and cell walls. The second pair, called stylets, are straws through which the leafhoppers slurp the liquids leaking from the wounded sites. To get more draw, they enlarge the breadth of the wound so more liquids seep out. Some species excrete digestive juices that cause cell collapse and the release of more sipables.

The wounded areas form a whitish scar pattern on the upper sides of the leaves, called stipling. When leaves are under severe stress they curl, develop yellow and brown spots around the wounded areas, and eventually fall from the plant.

Leafhoppers produce "honeydew," a sugary secretion composed of the sugary sap water from which the leafhoppers have filtered nutrients. Other members of their order Hemiptera such as aphids and scale also produce it, but leafhoppers are unique in their style. They produce bubbles of the viscous liquid from their posteriors, and when the bubbles reach the right size the leafhoppers release them to fall on other plant leaves or anything else below.

The balls of sweet syrup may attract ants, who may decide to herd some aphids on the verdant leaves. If the honeydew remains where it dropped, it may be colonized by sooty mold. This is harmful itself, but it also blocks valuable light from getting to the plant.

Leafhoppers are vectors for numerous bacterial, fungal, and viral diseases.

General Discussion

Leafhoppers are usually not much of a problem in home gardens, but given the right conditions, the perfect storm, they present a challenge.

Exclusion and Prevention

➤ Use row covers.

➤ Avoid overfeeding—high nitrogen encourages abundant tender growth, which is attractive to aphids.

➤ Inspect and quarantine all new indoor plant introductions.

➤ Leafhoppers are on the wing and looking for food and mates. Use a thrips filter, such as Horti-Control's Dust Shroom , or a HEPA filter in the air intake to exclude them from the ventilation airstream. Filter vents for the same reason.

CONTROLS

Azadirachtin	Insecticidal soap
Capsaicin	Neem oil
D-limonene	Pyrethrum
Garlic	Quarantine new plants
HEPA filter	Vacuuming
Herbal oils	Water spray
Horticultural oil	

BENEFICIAL BIOLOGICALS

Beauveria bassiana: The fungus is working when the leafhoppers turn reddish brown and have a fuzzy or shriveled texture.

Big-eyed bugs (*Geocoris spp.*)

Green lacewings (*Chrysopa fufilabris*)

Lady beetles (*Hippodamia convergens*), while effective, wander away in search of more plentiful food as soon as the prey population begins to drop.

Predatory midges (*Aphidoletes aphidimyza*)

Interesting Facts

Leafhoppers get very active as the temperature rises. For that reason, it is best to use control measures early in the morning when they are not as sprightly.

Leaf Miners

How common are these pests?

"Leaf miners" is a collective term for the larvae of a number of species of flies, a few moths, and even a species or two of beetles that "mine" or tunnel inside leaves. They are common in gardens.

What do they look like?

Larvae are light green to brown. Stubby maggots are about ⅛ inch long or less.

What kinds of plants do they attack?

Host preference depends on the species. Leaf miners are pests in vegetable crops, including beans, beets, chard, peas, potatoes, and spinach and ornamental plants,

Leaf miner: Orange-shouldered *Odontota scapularis*

shrubs, and trees as diverse as alder, azalea, birch, blueberry, bougainvillea, boxwood chrysanthemum, columbine, elm, holly lantana, nasturtiums, oak, and plantain.

Where are they found on the plant?

The larvae are found in the tunnels they create between the upper and lower surfaces of leaves.

What do they do to the plant?

Leaf miner damage is usually more aesthetic than an actual threat to the host. As the larvae feed, their bleached and yellowed tunnels scrawl across the leaves, making leafy vegetables undesirable to eat and ruining the appearance of ornamental plants. In some cases, leaf miners can destroy young seedlings, and on trees a severe infestation can weaken trees and make them susceptible to other disease organisms.

General Discussion

Most species of leaf miners overwinter as cocoons in the soil, although some over-winter as larvae that then pupate in spring. Adults emerge in early spring and lay small, white eggs on leaf surfaces. The resulting larvae burrow into the leaves and begin to tunnel their way through leaf tissue, leaving behind their characteristic winding tunnels.

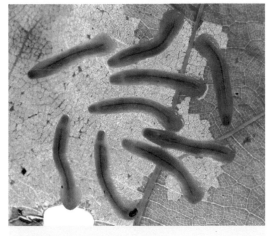

Leaf miner larvae

Once mature, some larvae cut their way through the leaf surface and drop to the ground where they pupate or tunnel into the soil to overwinter. Other species pupate within their tunnel.

Exclusion and Prevention

➤ Leaf miners are hard to control because during their most vulnerable stage—as lar-vae—they are safely tucked inside the tough surfaces of the leaves.

➤ Crush the larvae while inside their tunnels.

➤ Prune off affected foliage and destroy.

➤ Crush eggs found on the undersides of leaves.

➤ Keep weeds such as dock or lamb's quarters away from the garden as they are hosts for some species of leaf miners.

➤ Naturally occurring parasitic wasps are perhaps the best control for leaf miners as they know how to get at their prey. Attract parasitic wasps by planting nectar-producing plants in and around the garden.

➤ Use floating row covers.

➤ Rotate crops. Avoid susceptible species of plants.

➤ Plant late.

➤ Keep plants and trees healthy so they can tolerate an infestation.

CONTROLS

Azadirachtin	Neem oil
Capsaicin	Pyrethrum
Herbal oils	Spinosad
Horticultural oil	

BENEFICIAL BIOLOGICALS

Beauveria bassiana	**Wasps, parasitoid** (Native)

Leaf miner trails on leaf

Interesting Facts

Entomologists can tell which type and species of leaf miners are present in a leaf by studying the tunneling pattern they leave behind. Serpentine leaf miners leave behind winding, snakelike patterns that get wider as the larvae grown. Blotch leaf miners leave behind irregular, rounded areas. Tentiform leaf miners leave behind bulging, tent-shaped knobs on leaf surfaces that turn brown and dry out.

Mealybugs

How common are these pests?

Mealybugs are a common pest on both indoor and outdoor plants. There are about 280 species found in North America. They are closely related to scale and are a member of the order Hemiptera, the true bugs, which have mouths modified to suck plant juices.

Longtailed mealybug (*Pseudococcus longispinus*)

What do they look like?

Adult female mealybugs are flat, oval, pink, segmented, and generally covered with waxy white fluff. Most species are about ⅛ inch long. Nymphs are similar to adults but smaller. Males are tiny, two-winged flies that do not feed and are rarely seen.

What kinds of plants do they attack?

Favorite hosts include citrus and fruit crops, such as apples, avocadoes, grapes, lemons, oranges, and pears, and flowers, orchids, ornamentals, potatoes, and tropical foliage plants such as the *Dieffenbachia* genus, dracaena, and hibiscus.

Where are they found on the plant?

Mealybugs are found on all parts of host plants, but they prefer the nodes of leaves and stems, crevices, and new growth.

What do they do to the plant?

Mealybugs have piercing/sucking mouthparts and feed on plant juices. Heavy feeding results in yellowing, withering, and premature fruit drop. Mealybugs also produce honeydew, a rich sugar-laden excrement that ants love and that provides an excellent environment for sooty mold to grow. Once mealybugs are established, ants often move in to "farm" them, as they do aphids, to obtain the sweet honeydew.

General Discussion

Females lay eggs in batches of 200 to 1,000 on fruit, leaves, and twigs. Some species encase their eggs in fluffy white egg sacs or shelter them either on or underneath their bodies. The eggs hatch in one to four weeks and the .04 inch (1 mm) nymphs then begin to feed in groups for one to two months until they mature into adults. Several generations a year are possible outdoors.

Citrus mealybug (*Planococcus citri*)

Exclusion and Prevention

➤ Carefully inspect any new plants brought into the garden or greenhouse for hitchhiking mealybugs.

CONTROLS

Alcohol spray	Neem oil
Azadirachtin	Pyrethrum
Citrus oil	Remove mealybugs from plants with a cotton swab or cotton ball dipped in **rubbing alcohol**.
Herbal oils	
Horticultural oil (especially sesame oil)	
Insecticidal soap	Use a strong stream of **water** to knock mealybugs off plants.

BENEFICIAL BIOLOGICALS

Beauvaria bassiana	**Mealybug destroyer** *(Cryptolaemus montrouzieri)* for grape, citrus, or indoor plantings
Bacillus thuringiensis var. *kurstaki*	
Hoverflies	**Minute pirate bugs**
Lacewings	**Wasps, parasitoid** *(Leptomastix dactylopii)* for citrus crops
Ladybeetles *(Ryzobius lophanthae)*	

Interesting Facts

Mealybugs have a relatively long reproductive cycle so they take a long time to reach infestation numbers. To eliminate them, treat the plants three or four times two weeks apart. This interrupts their life cycle.

Moles

How common are these pests?

Moles are common in temperate rural areas where the soil has been loosened by cultivation, less so in urban areas.

What do they look like?

Moles are burrowing mammals of the *Talpidae* family, not rodents. They grow 5 to 7 inches in length and weigh around 4 ounces when adults. They have dark, soft fur, small eyes, pointed

Mole

snouts, and powerful front digging claws. They are rarely seen by gardeners as they primarily remain below ground but are most often detected by the raised surface of their burrows that zigzag across garden spaces.

What kinds of plants do they attack?

Moles rarely eat plants, feeding instead on insects, earthworms, and gophers in the soil. They do more good than harm but can be obtrusive in the garden or lawn.

Where are they found on the plant?

Moles build complex tunnel systems in rich, moist soil where they can find an abundance of insects.

What do they do to the plant?

Moles seldom directly damage plants. Instead, their tunnels and mounds disturb the root zone of plants, allowing plant roots to dry out and causing unsightly bumps and mounds in lawns and hazards for walkers.

General Discussion

Moles can be easily distinguished from gophers by the shape of their digging. Moles create rough cones with a hold or earthen plug near the center. Gopher mounds are fan shaped with the hold or plug near the narrow end.

Moles have a single litter of two to five pups a year. They birth in mid- to late spring. The pups' survival rate is around 50 percent. Except when breeding, moles are solitary animals and defend their tunnel system to the death when invaded by another mole.

Exclusion and Prevention

➤ Line planting holes on the bottom and sides with hardware cloth.

➤ Castor oil is effective as a repellent and is sold under a number of brands and names.

➤ Use predator urine.

➤ Electronic devices that emit irritating vibrations through the soil are sold widely as mole repellents, but their effectiveness is anecdotal.

CONTROLS

➤ Since moles do little damage directly to plants, there is no need to eradicate them. However, if it does become necessary to remove them, trapping and fumigating are the best methods. Commercial fumigants sold for use against gophers are also effective for moles.

Mole traps are available as the lethal or live-trap sort.

Carbon dioxide from dry ice or a tank is effective as a fumigant.

BENEFICIAL BIOLOGICALS

None

Interesting Facts

An adult mole weighing around 5 ounces eats 45 to 50 pounds of insects and earthworms each year.

Moles can dig tunnels at the rate of 18 feet per hour and move through their existing tunnels at about 80 feet per minute.

Moles have twice as much blood with twice as much hemoglobin as other mammals, allowing them to breathe more easily in their low-oxygen underground environment.

Mole hole

Nematodes, Foliar

How common are these pests?

These tiny roundworms, genus *Aphelenchoides*, are common in all garden areas when the right hosts are available. Foliar nematodes spend their lives either in living tissue or overwintering in plant debris in the soil.

Foliar nematode leaf damage

What do they look like?

Foliar nematodes are roundworms that look like little worms but you will probably never see them because their entire life cycle takes place inside the leaf epidermis.

The nematode mouth is radially symmetrical so its six lips form a circle with teeth on all sides that chew its prey. The head is distinct from the rest of the body. Adults are less than 1/8 of an inch (2.5 mm) long.

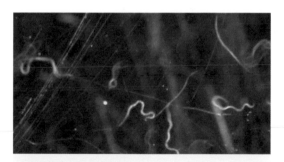

Adult foliar nematodes

What kinds of plants do they attack?

Foliar nematodes attack a wide variety of plants, but particularly susceptible plants include African violets, asters, azaleas, carnations, chrysanthemums, coneflowers, dahlias, ferns, hostas, phlox, primroses, rhododendrons, and strawberries.

Strawberry crimp nematode damage

What do they do to the plant?

Foliar nematodes damage the flowers, leaves, and shoots of susceptible plants. Damage from foliar nematodes starts in leaves near the bottom of plants and moves upward. Leaves show V-shaped chlorotic patches that are usually confined by the leaf veins—a classic diagnostic tool for foliar nematodes. These areas widen, turn black or brown, and eventually fall out, leaving holes in the leaves. Shoots and buds from plants infested, but not killed, produce twisted, distorted, and stunted growth.

To see the worms, place leaf tissue in a container with water for a day. The tiny wiggly worms appear in the water and can be seen using a magnifying lens.

Exclusion and Prevention

- Avoid susceptible types of plants.

- Remove and destroy infested plants or foliage. Do not compost them

- Space plants to allow leaves to dry between rains or watering. Drip irrigation minimizes wetness, impeding nematode transfer between plants.

- Keep growing areas as dry as is practical. Foliar nematodes move about on host plants and between plants on a thin film of water.

Nematodes infecting larva host

- Foliar nematodes survive for long periods in an inactive state in dried plant material, especially under cool conditions. Remove garden debris as this provides overwintering sites.

- Quarantine newly acquired susceptible plants and look for symptoms before introducing them to an established growing area.

CONTROLS

Azadirachtin—however, it also kills beneficial nematodes

Compost tea

Hydrogen peroxide

Insecticidal soap

Quarantine new plants

Dip infested plants into **120°F water** for 10 minutes followed by a plunge into cold tap water as plants are breaking dormancy.

BENEFICIAL BIOLOGICALS

Trichoderma harzianum

Nematodes, Root Knot (*Meloidogyne* spp.)

How common are these pests?

Root knot nematodes are common in most garden soils, but they are especially damaging to crops in sandy soils.

What do they look like?

Root knot nematodes are small round worms with a radially symmetrical mouth, a head, and the rest of the body is thin and unsegmented. They have six jaws with teeth, arranged in a circle. They are not visible to the naked eye but can be seen using a photographer's loupe. The knots that they create on roots where they live in large colonies are more apparent than the nematodes.

What kinds of plants do they attack?

Root knot nematodes attack a wide variety of plants, more than 2,000 separate species, but are most harmful to many garden and agricultural crops, including bean, carrot, fruit trees, lettuce, melon, onion, pea, pepper, potato, sweet potato, and tomato.

Root knot nematodes on coleus

Where are they found on the plant?

Root knot nematodes are found in soils in hot climates or short winters. The juveniles are found in the soil and attack the roots of plants.

Root knot nematodes formed galls or knots on these pumpkin plant roots

What do they do to the plant?

Aboveground symptoms include yellowing, wilting, stunting, and death. Fruit or flowers fail to form or are stunted. Roots infested with root knot nematodes form large and small galls that compromise the plant's ability to absorb water and nutrients. Females lay eggs and juveniles feed inside the galls. Root knot infection also weakens plants, putting them at risk of attack by other pathogens.

General Discussion

Plant resistant or tolerant crops or cultivars. Plants with resistance are marked with the initials VFN. V stands for resistance to Verticillium wilt, F stands for resistance to *Fusarium* wilt, and N represents resistance to root knot nematodes. Resistant varieties suppress the nematode populations. Tolerant varieties grow in infested soil, do not suppress the populations, but produce yields. There are resistant varieties of many flowers and vegetables. If root knot nematodes are a problem in your area or if your garden has suffered from root knot nematodes in the past look for the VFN mark.

Varieties that are resistant to one population of nematodes may not be to another. At the end of the growing season, pull up some plants and look for galls. If root knot nematodes are feeding on these plants, then try another variety with resistance.

Grow early-season vegetables such as bean, English pea, lettuce, pea, potato, radish, and sweet potato. These may be grown in infested soil with limited damage.

Exclusion and Prevention

▶ Rotate susceptible vegetable crops with crops that are less susceptible such as asparagus, fescue, marigold, onion, or small grains. This deprives the nematodes of food and reduces the population. Grow resistant crops in areas infested with root knot nematodes for three consecutive years.

▶ Remove crop roots immediately after harvest.

▶ Turn the garden soil two to four times

Galls or nodules caused by root-knot nematodes.

in fall to expose the nematodes to weathering and predation.

- Keep the areas in and around the garden weed free as some weeds species are hosts for root knot nematodes.

- Avoid spreading infested soil to non-infested areas.

- Apply *mycorrhizae* and *trichoderma* to planting mix.

- Incorporate compost and humus into the garden soil to help promote beneficial organisms that naturally control nematodes and to increase the moisture-holding capabilities of the soil.

CONTROLS

Pepper plant roots with extensive galling, heavily infested by the southern root-knot nematode.

Azadirachtin—however, it also kills beneficial nematodes.

Fallowing—Allow infested parts of the garden area to lie fallow or unplanted for several gardening season to starve the nematodes and reduce populations.

Marigolds—Plant a marigold cover crop, especially African, French, or Mexican types, in and around the garden. French varieties Petite Harmony and Petite Gold reduce nematode populations. Avoid hybrid varieties as they may actually increase the numbers of nematodes.

Soil Solarization—Solarize areas of the garden infested with nematodes.

BENEFICIAL BIOLOGICALS

Nematodes, predatory (Steinernema feltiae)

Rats

How common are these pests?

Rats are often found in gardens in suburban or urban areas, in gardens close to buildings, or in dense areas of vegetation.

Black rat adults are 6 to 12 inches long, excluding tail, weigh 7 to 14 ounces, and prefer dry upper stories.

What do they look like?

Rats are rodents ranging from 10 to 16 inches in length, depending on the species, not including their tails. They weigh from 6 to 12 ounces and have dark fur ranging from brown to black. Their heads are long and taper to a snout with long whiskers, and they have round, prominent ears.

What kinds of plants do they attack?

Rats eat fruits and nuts and sample other garden produce, such as corn and squash. They also chew on or burrow underneath ornamental plants.

Where are they found?

Rats have followed humans around the world. Europeans brought both the brown and black rat to North America. They live near you, although they may not always visible. Their presence is most often detected by the trail of droppings they leave behind.

What do they do to the plant?

Rats do most of their damage to gardens by gnawing plants and leaving them to die, or else by burrowing around the plants until the root systems are compromised and die. Rats prefer to eat berries, corn, fruit, and nuts, but all food scraps left in the garden draw rats. They also feast on snails and some slugs as well as earthworms and large insects. They are drawn by animal odors such as blood meal. They burrow, dig, tear, and destroy to get to the food, so it is best to clean up after yourself and to keep the garden free of debris that might be eaten.

General Discussion

Rats are prolific breeders. Females can produce four to nine litters per year depending on food availability, usually with five to 12 babies per litter. They adjust their fertility to suit the food supply. In wild areas rats eat other small animals, fruits, grains, insects, nuts, seeds, and other organic matter.

Rats damage garden plants and cause other problems, including contaminating foods or grains with their urine or droppings, chewing house insulation and wiring, carrying parasites and diseases, biting or killing small house pets, and biting people.

Exclusion and Prevention

➡ Clear trash and brush from near the garden area. Plant as far away from attractive food plants such as fruit trees or berry bushes and away from barns or other buildings that can harbor nests.

➡ Feed outdoor cats in the garden area. Rats avoid areas that smell of cat urine.

➡ Create barriers by wrapping chicken wire or hardware cloth around the plants 18 inches high. Steel wool wrapped around the stems of plants and secured with twist ties stops stem chewing.

➡ Install rat fencing.

➡ Plug tunnels and entrance holes using steel wool.

➡ Tightly cover or protect all possible food sources.

Brown rat adults are up to 10 inches long, weigh up to 12 ounces, and prefer ground-level environments.

CONTROLS

Flooding

Fumigation

Rat traps. Use peanut butter as bait.

Poisons: Two new safe poisons are cholecalciferol (vitamin D3) and zinc phosphide. Place rodent bait in tamper-proof stations. Rats have developed resistance to the classic anticoagulant rat poison warfarin, which is unsafe to use around pets and is toxic to wild predators that feed on the dead rats.

BENEFICIAL BIOLOGICALS

Birds of prey   Dogs

Cats   Snakes

Interesting Facts

Rats are among the most successful mammals. They have populated the world while allowing humans to make the travel arrangements. They just hop aboard for the free lunch.

Two species of rats, the brown and the black, dominate. See the chart for more information on them.

Comparison of Characteristics of Black and Brown Rat		
	Black Rat	**Brown Rat**
Color	Predominately black with some individuals brown to gray. Underbelly gray to white	Brown upper body with brownish underneath
Size	Slim in build, weighing 5 to 10 ounces	Large, weighing 12 to 19 ounces
Tail	As long as body with fine, black scales	Shorter than body, dark upper and pale underneath with scales
Muzzle	Pointed	Blunt
Ears	Long, will reach eyes if folded over	Short, will not reach eyes
Habitat	Burrows along buildings, beneath rubbish, in woodpiles. Indoors, remain in basements	Live and nest aboveground in trees, shrubs, and other dense vegetation. Indoors, remain in attic spaces, walls, false ceilings, and cabinets
Droppings	Larger than mouse droppings, thin	Large, oval, larger than black rat

Root Aphids

How common are these pests?

Root aphids occur occasionally in indoor gardens and in greenhouses all over the world.

What do they look like?

Root aphids look very similar to common aphids with the exception that their cornicles, which look like dual tailpipes on most aphids, are shortened and curved in the soil dwellers.

Root aphids

What kinds of plants do they attack?

These aphids attack the roots of many vegetables and ornamentals.

Where are they found on the plant?

Root aphids attach their stylets to tender roots in order to suck their juices.

What do they do to the plant?

As the population grows and the aphids' consumption increases, the plant's vitality is sucked away until it gradually fades away.

General Discussion

Root aphids are a new and troubling phenomenon for indoor and greenhouse gardeners. Sometimes when they show up in gardens, there is no apparent indication as to the cause of infection.

Exclusion and Prevention

► Use only sterile planting medium or pasteurized soil in your container plants. With sensitive indoor plants from friends, presume that they are infected and if appropriate, start from cutting or seeds rather than rooted clones. Quarantine all new plants for at least a week. Pull plant away from pot and inspect roots. If damaged with scars or wounds, this is an indication of pests.

► Root pests have appeared in extremely well-kept isolated gardens. Filtering water with a UVC light will knock out juvenile insects traveling in the water.

Root aphids

CONTROLS

D-limonene	Pyrethrum
Herbal oils	Quarantine new plants
Neem oil	Spinosad

BENEFICIAL BIOLOGICALS

Beauvaria bassiana	Mites *(Hypoaspis miles)*
Bt-k	Nematodes, beneficial

Interesting Facts

Root aphids were absent or a rare occurrence in the 1990s. Since then they have become more common.

Sawflies

How common are these pests?

Many species of sawflies are distributed throughout North America.

What do they look like?

Adults sawflies are wasplike insects that are rarely seen. Their name comes from the sawlike ovipositor that adult females use to cut slits in host plant tissue and deposit their eggs. Adults are from ¼ to ¾ inch long with clear

Sawfly larvae on leaf undersides

wings. Despite their similarity to wasps, they do not sting. The larvae can vary widely in appearance among the many species. Some may look like caterpillars, except that sawfly larvae have three pairs of true legs and six pairs of prolegs, as compared with caterpillars, which have five or fewer pairs of prolegs. The larvae of some species may look like slugs. Larvae color varies from green and gray to greenish yellow, and they may have stripes or spots.

What kinds of plants do they attack?

Sawflies are divided into groups depending on the plants they frequent: evergreens and deciduous plants. Evergreen hosts include pines and other conifers such as arborvitae, hemlocks, and junipers. Deciduous plant hosts include flowers, shrubs, and trees, including azaleas, birch, columbine, cotoneaster, currants, fruit trees, gooseberry, hawthorn, raspberry, roses, willows, and other plants.

Where are they found on the plant?

Sawfly larvae feed on the upper leaves or needles of plants.

What do they do to the plant?

Adult sawflies do not harm plants, but the larvae feed in large groups and can strip a plant by skeletonizing the leaves and causing premature leaf drop, or by consuming whole leaves. On conifers, they consume the older foliage before the spring bud break, creating a "bottle brush" appearance to the now-stripped branches when the new apical growth emerges. Once a branch or limb has been stripped of food, the larvae move on to new feeding places. Some species of sawflies can completely defoliate a plant; others only cause aesthetic damage.

Sawfly damage

General Discussion

Most species of sawflies overwinter as pupae or prepupae in the soil and emerge as adults in spring. Females insert eggs in needles or leaf tissue by cutting tiny slits with their ovipositor. When the eggs hatch in early summer, the larvae feed through summer while undergoing several molts before dropping to the ground to pupate. Some sawfly species emerge as adults in fall and lay their eggs, which overwinter. Depending on the species, several generations may be produced per year.

Exclusion and Prevention

Elm Sawfly larva

➥ For small infestations, handpick the larvae when they are small.

➥ Spray the plant with a forceful stream of water to knock the larvae to the ground.

➥ Shake plants to dislodge larvae. Catch in a drop cloth and destroy by crushing or dropping in soapy water.

➥ Remove branches or limbs infested with large groups of larvae and destroy.

➥ Hang white cards coated with nondrying glue in the tops of host plants to trap adults.

- Plant resistant varieties.

- Plant susceptiblc pine varieties in-termixed with deciduous trees.

- Keep susceptible plants healthy.

- Attract birds that prey on saw-fly larvae, such as chickadees and wrens, to the garden by providing water and a nesting habitat.

Adult sawfly

CONTROLS

Horticultural oils	Pyrethrum
Insecticidal soap	Spinosad
Neem oil	

BENEFICIAL BIOLOGICALS

Beauvaria bassiana	Tachinid flies
Damsel bugs	Wasps, parasitoid

Interesting Facts

Bt *(Bacillus thuringiensis)* will not control sawfly larvae.

The larvae of conifer sawflies display interesting defense mechanisms such as rearing up when a hand is passed over them, mass twitching in the presence of a perceived threat, or disgorging salivary fluids.

Scale

How common are these pests?

More than a thousand species of scale occur throughout North America. They are common in greenhouses and outdoors.

What do they look like?

Scale are generally divided into two groups: soft scale and armored scale. Female soft scale are oval and generally brown to gray, legless, and wingless, and are about 1/10 to 1/5 inch in diameter. They appear on leaves and stems as soft bumps. Males are tiny, yellow, winged insects that are rarely seen. Immature forms are called "crawlers" and resemble mealybugs in appearance.

Armored scale are circular or oval, hard shelled, and about 1/10 inch in diameter. Some species have a small dimple in their back. They secrete their armor of wax in an oyster-shell pattern. This armor is not a part of their bodies; instead, they use it as protection while they feed underneath,

Soft Scale (*Coccidae*)

Armored Scale (*Diaspididae*) have a platelike cover

hence their name. Colors vary from white, yellow, and gray to reddish brown and purplish brown. Early-stage nymphs are mobile and resemble soft scale crawlers. As they mature, the females lose their legs and become sedentary.

What kinds of plants do they attack?

Host preference varies by species, but scale attack citrus, fruit trees, grapes, houseplants, orchids, palms, roses, and shrubs. Perennials are more likely to suffer damage because they have a slow life cycle, taking more than a month to become reproductive.

Where are they found on the plant?

Scale are found on the twigs and leaves of plants.

What do they do to the plant?

Scale are true bugs with piercing/sucking mouthparts perfected to feed on plant sap. Some species inject toxic saliva into the plant tissues to prevent the plant's natural defenses from closing up wounds. The first signs of damage are tiny dots or blotches on the leaves. In severe infestations, leaves yellow and drop and the plant becomes stunted or dies. Scale produce honeydew, the sugar-rich excrement that attracts ants and serves as a media for sooty mold to grow.

General Discussion

Depending on the species, females lay eggs or give birth to live young. Nymphs wander around on leaf surfaces until they find a feeding spot and settle in. Armored scale may take a month or more to molt into an immobile adult form. For soft scale, females molt into a nearly legless, immobile form. Males molt into winged, gnat-like insects that fly around searching for mates.

Exclusion and Prevention

➨ Wipe scale off plants with a cotton ball or cotton swab dipped in rubbing alcohol.

➨ Prune and dispose of affected branches.

➨ Use resistant varieties.

CONTROLS

Alcohol spray	Herbal oils
Azadirachtin	Insecticidal soap (soft scale)
Dormant season horticultural oils (both types of scale)	Neem oil (soft scale)
	Pyrethrum (soft scale)
Growing season horticultural oil (soft scale)	

BENEFICIAL BIOLOGICALS

Beetles, predator (*Chilocorus nigritus* or *Lindorus lophanthae*)	Ladybeetles
Lacewings	Wasps, parasitoid (*Metaphycus helvolus* for soft brown scale and *Aphytis melinus* for oleander scale)

Interesting Facts

When scouting the garden for presence of scale, be aware of tiny, pin-size holes in any armored scale's hard shell. These are evidence that tiny parasitic wasps are already at work preying on the scale.

Slugs and Snails

How common are these pests?

Slugs and snails occur throughout North America.

What do they look like?

Slugs are soft-bodied, shell-less mollusks with a pair of prominent tentacles on their head. Snails appear much the same except that they are enclosed in hard coiled shells that they recede into when threatened. Both move by way of a muscular "foot" on the undersides of

European brown snail

their bodies. Slugs and snails leave a telltale trail of glistening, thick mucus behind as they crawl over surfaces. Color ranges from brown to black to gray, for most common species. Eggs are round, clear, and laid in jellylike masses underneath debris and stones.

What kinds of plants do they attack?

Some species of snail eat nearly every garden plant, including flowers, ornamental shrubs, trees, and vegetable seedlings. They do avoid plants with fuzzy surfaced leaves or those with strong odors such as herbs and lamb's ears.

Where are they found on the plant?

Slugs and snails prefer young seedlings but will also eat stems, leaves, and ripening fruit. Both feed primarily at night.

What do they do to the plant?

Slugs and snails chew irregular holes on leaf margins or centers by rasping the surface of the leaves with fine teeth and leaving behind trails of mucus. Several young, tender seedlings can be devoured in a single night. They do the most damage in moist, shaded locations.

General Discussion

Slugs and snails are hermaphroditic and can fertilize themselves if they cannot find mates. Females lay eggs in clutches of 30 to 120, 1 to 2 inches deep in moist soil, or hidden beneath boards, debris, or stones. Eggs hatch, and the young crawl away and begin to feed. Some snails take up to two years to mature and form their shell. In dry conditions, slugs and snails can go dormant until conditions improve.

Exclusion and Prevention

► Create a barrier with iron phosphate. Iron phosphate acts as a poison to snails and slugs, and once they consume it, they will stop eating and soon die. Iron phosphate is harmless to plants, humans, and pets.

► Remove garden debris to eliminate overwintering and egg-laying sites.

► Water in the morning to allow the garden to dry out before evening.

► Handpick at night. Hunt with a flashlight because they feed after dark. Look under boards, flowerpots, or pieces of potato or cabbage leaves placed in the garden.

Decollate snails *(Rumina decollate)* prey on the eggs and body of small to medium sized brown snails. Native to the Mediterranean, this snail was introduced into the United States as a biological control and has since spread along the sun belt from California to Florida. They can also be bought commercially. This snail has a distinct cone-shaped shell in contrast to the simple round spiral of other snails. If found in the garden with other snails, leave them to hunt. But if they are the only snails in the garden, they are likely feasting on the garden plants, so go ahead and remove them. They are illegal to use in some areas so check first.

Decollate Snail

Gray garden slug

- Trap in jars of beer or sugar water and yeast. They are attracted to the odor and drop into the brew.

- Wrap copper strips, tape, or mesh around the garden perimeter or around individual plants. The copper delivers a mild shock that the crawlers avoid.

- Sprinkle diatomaceous earth at the base of plants to deter slugs and snails. However, it will also deter some beneficial insects.

Copper tape

- Raise plants above their crawl level

- Create a shaded damp area that they are drawn to during the heat of the day. Then raid it.

- Coffee grounds, coarse sand, and wood ashes sprinkled around plants will also act as deterrents.

CONTROLS

Iron phosphate bait

BENEFICIAL BIOLOGICALS

Free-ranging poultry such as chickens, ducks, and geese gobble up snails and slugs.

Encourage **frogs, salamanders, toads,** and **turtles** to visit the garden by adding a pond to your yard. Slugs are one of their favorite snacks.

Nematodes, predatory
(*Phasmarhabditis hermaphrodita*)

Interesting Facts

That infamous slime that slugs and snails leave behind is really a complex polymeric compound that absorbs water at 100 times its weight allowing the "foot" to easily glide over the lubricated surfaces.

Spider Mites

How common are these pests?

Spider mites are common pests outdoors and indoors.

What do they look like?

Spider mites are tiny. Most species are less than 0.4 mm long and visible only with the use of a hand lens. With four pairs of legs and no antennae, they are like their relatives spiders and ticks, but unlike spiders, spider mites have only one body segment. Colors vary from red to brown, black to yellow, depending on the species and the season. Color also varies within species, making it an unreliable tool for identification. Fortunately for gardeners, spider mites tend to live in large groups and spin fine webbing for protection, which gives away their location.

What kinds of plants do they attack?

The host range for spider mites as a group is very large and includes berries, conifers, fruit trees houseplants, ornamental shrubs, shade trees, and vegetables. Particularly attractive to spider mites are apples, apricots, beans, blackberries, black currants, chrysanthemums, gooseberries, grapes, hops,

Red spider mite

Two-spotted spider mite

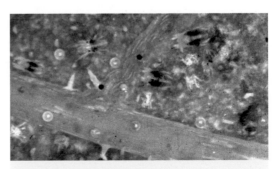
Two-spotted spider mite life cycle

kiwifruit, nectarines, peaches, pears, peas, raspberries, roses, strawberries, sweet corn, and walnuts.

Spider mite webbing

Where are they found on the plant?

Spider mites feed primarily on the undersides of leaves and also the stems of plants. They can also be found in the buds and flowers.

What do they do to the plant?

Spider mites pierce the surface of the leaves, then suck plant juices from them. These punctures appear on the leaves as tiny brown spots surrounded by yellowing leaf. They can be seen as colored dots on the leaf undersides. As the population grows they produce webbing that the mites use as a pedestrian bridge between branches and plants. Spider mites are also vectors for disease, since they travel from leaf to leaf and to other plants.

General Discussion

When you see signs of infestation, take action immediately. Spider mites have a high rate of reproduction and are among the most feared of garden pests. Population ratios for spider mites are three to one, female to male. Females can lay up to 200 eggs, and the life cycle can repeat as often as every eight days in warm, dry conditions.

Spider mites are by far the most fearsome of all plant pests. They suck plant juices, weakening the plants. They multiply quickly. Spider mites are most active in warmer climates than cold ones.

Exclusion and Prevention

➤ Look for the yellow-brown spots mites leave when feeding and for their webs. Infected plants transmit mites. When you spot mite symptoms, take action immediately.

➤ Inspect all plants introduced into the garden. Place them in quarantine for 10 days, or treat all introductions as a precaution.

➤ Knock off mites with a strong stream of water.

➤ Plant varieties less susceptible to spider mites.

➤ Spider mites thrive in dry climates. High humidity slows spider mite development and reproduction.

➤ Filter incoming air with a HEPA filter

CONTROLS

Azadirachtin

Capsaicin

Cinnamon-clove tea: Use an ounce of each spice per gallon of water. Boil water and let stand a couple of minutes, add powdered spices, and let brew until cool. Strain before spraying.

Garlic

Herbal oils (cinnamon, clove, coriander, D-limonene, and rosemary)

Horticultural oils (dormant season) sprayed in late winter or early spring will kill overwintering eggs.

Insecticidal soap

Neem oil

Pyrethrum works with some populations. Others are immune to it.

Quarantine new plants

Spinosad

Sulfur burner

BENEFICIAL CONTROLS

Beauveria bassiana (beneficial fungi)

Big-eyed bugs

Damsel bugs

Lacewings

Lady beetles

Minute pirate bugs

Mites, predatory: Get the best variety suited to your garden and introduce early. Control may still be challenging. Predator mite species vary in the environmental conditions they require and in their eating habits.

Interesting Facts

The two-spotted spider mite *(Tetranychus urticae)* is the most damaging spider mite worldwide. Populations have become immune to most miticides.

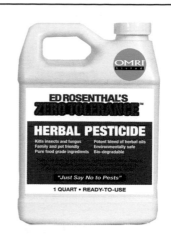

Eliminate mites by breaking up the life cycle. Spray Ed Rosenthal's Zero Tolerance three times, three to four days apart.

Microscopic view of two-spotted spider mite

Mites congregating

Squash Bugs
(Anasa tristis)

How common are these pests?

The squash bug is common in gardens growing cucurbits such as cucumber and squash.

What do they look like?

Adults are grayish brown and about ⅝ inch long with oval, flat backs and are covered with fine, dark hairs. The head is prominent, and the edge of the abdomen is often copper colored or striped. Newly hatched nymphs are green and wingless. As they age, nymphs turn gray and develop prominent wing pads.

What kinds of plants do they attack?

All cucurbit crops are susceptible such as cucumbers, pumpkins, squash, and zucchini. Melon and summer squash are less susceptible than some other plants. Winter squash is often the most severely affected.

Where are they found on the plant?

Adults and nymphs suck the juices from the stems and leaves and also feed

Late season mass of squash bugs on pumpkin

Squash bug eggs

on the fruit. Large numbers of individuals can be found around the stems and fruits and even on the ground underneath susceptible plants.

What do they do to the plant?

When squash bugs suck the juices from leaves and stems, they inject a toxin that results in plant wilt, yellowing, and death of both the leaves and the plant beyond the visible damage. Damage is first noticeable as small yellow flecks on leaves that later turn brown.

General Discussion

Unmated adults overwinter in protected areas around last year's feeding sites under boards, leaf debris, and stones. They emerge in late spring and mate, and females lay large groups of eggs on the undersides of leaves, usually in the vein axis and on stems. Eggs are oval and yellow when first laid, then turn orangish brown close to hatching. They hatch in 10 to 14 days. Nymphs feed in groups and go through five molts before emerging as adults in late summer.

Exclusion and Prevention

➤ If only a few plants are affected, hand-pick squash bugs and eggs.

➤ Trap them under boards or shingles placed on the ground near the plants. At night squash bugs congregate under the boards. Remove them each morning.

➤ Floating row covers. Remove the covers for pollination or hand-pollinate.

➤ Vacuum adults and larvae.

Wilting associated with squash bug infestation.

➤ Look for and destroy the eggs masses.

➤ Remove garden debris to eliminate overwintering sites.

➤ Interplant susceptible plants with deterrent plants like dill or marigolds.

➤ Trellis plants to keep the stems and vines off the ground.

➤ Plant pollen and nectar plants in the garden to attract parasitic flies.

➤ Plant resistant varieties of cucurbits such as acorn, blue Hubbard, butternut, spaghetti, and zucchini squashes.

➤ Rotate susceptible crops with noncucurbit crops.

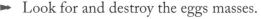

CONTROLS

Azadirachtin

Herbal oils

Insecticidal soap

Neem oil

Pyrethrum

Spinosad

BENEFICIAL BIOLOGICALS

Beauveria bassiana

Praying mantids

Tachinid flies *(Trichopoda pennipes)*

Older nymphs on sqaush

Interesting Facts

Squash bugs emit a foul odor when crushed or when they are present in large groups.

Thrips

How common are these pests?

Thrips are common garden pests.

What do they look like?

Thrips are less than 0.06 inch long but can still be seen by the naked eye. Colors range from yellow to dark brown. Adults are winged but are poor fliers and jump when startled. The head and body colors range from yellow to dark brown. Larvae are about one-half the size of the adults and are lighter in color and wingless.

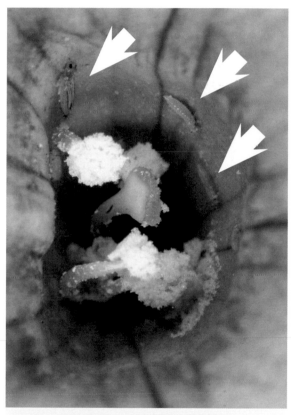

Western Flower Thrip (*Frankliniella occidentalis*)

What kinds of plants do they attack?

Thrips, depending on the species, feed on a wide variety of plants, including avocado, azalea, beans and other legumes, blueberries, citrus, dogwood, garlic, gladiolas, grapes, hops, impatiens, iris, melons, onions, pears, pepper, petunia, privets, shrubs, roses, squash, stone fruits, strawberries, tomatoes, and other plants both indoors and out.

Where are they found on the plant?

Thrips feed on the leaves of plants and are usually found on the top surface.

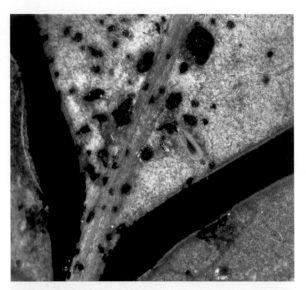
Wester Flower Thrip damage

Western Flower Thrip

What do they do to the plant?

Thrips use an unusual saw-like mouthpart to scrape and pierce the leaf surface until the sap begins to flow. Then they suck up the juices, leaving the leaf surface stippled with white or silvery scrapes or scars. Leaves eventually turn whitish in color. Thrips leave behind greenish-black specks of feces on both the upper and lower leaf surfaces. Thrips damage may also be mistaken for spider mite or leaf miner damage at first, but more severe cases result in color-stripped leaves.

Severely damaged leaves will not recover, and their ability to absorb light is compromised. If thrips are not controlled, the plants die. Thrips also carry pathogens that they transfer from plant to plant.

General Discussion

Outdoors, thrips hibernate over winter in soil and plant debris. Thrips become active when the temperature climbs above 60°F (16°C). The warm, stable temperatures of indoor gardens allow them to be active year-round. Thrips are a more serious problem indoors because of this, and also because a natural soil-dwelling fungus that infects thrips pupae is not present indoors.

Females lay eggs (anywhere from 40 to 300 depending on species) in plant crevices or actually insert them into the leaves and stems. The larvae feed until they enter the pupal stage, when they fall to the ground (and when the soil fungi provide some biocontrol outdoors). Depending on the species and temperature (optimum is 77°F to 82°F [26°C to 28°C]), the larval thrips hatch, pupate, and mature to egg-laying adults in seven to 30 days.

Exclusion and Prevention

➤ Thrips pupae live in the soil after they drop from the plant. A barrier placed over the soil prevents pupae from getting to the plant and they die. As with fungus gnat larvae, a layer of dry diatomaceous earth on top of the soil also helps to destroy the thrips pupae.

➤ Thrips are drawn to the colors pink, blue, and yellow, so it's best to avoid having yellow walls or items around your garden. Use blue, pink, or yellow sticky cards as indicator traps to detect an infestation of thrips. Use garlic in outdoor gardens to deter or repel thrips.

➤ Place a barrier on the top of the soil of indoor plants to block pupae from reaching the soil. Paper, cardboard, co-comats all work.

The Dust Shroom from Horti-Control keeps thrips, mites, molds, and harmful bacteria from entering indoor garden spaces.

CONTROLS

Azadirachtin

Capsaicin

HEPA Filter

Herbal oils (cinnamon, clove, and coriander)

Horticultural oil

Insecticidal soaps

Neem oil

Pyrethrum

Spinosad

BENEFICIAL BIOLOGICALS

Beauveria bassiana (beneficial fungi)

Minute pirate bugs (*Orius*) attack adult thrips.

Mites (*Hypoaspis miles*)

Nematodes, beneficial (*Heterorhabditis bacteriophora and Steinernema feltiae*)

Whiteflies

How common are these pests?

Whiteflies are found throughout North America and are a common pest both indoors and out.

What do they look like?

They are 0.04 inch (1 millimeter) long, and their soft bodies are covered in a powdery wax that gives them protection and their white color. They are related to aphids, leafhoppers, and scales. When foliage is disturbed, they flutter up in profusion. Immature stages are mostly immobile and scale-like while feeding. Eggs are pinpoint size, yellow or gray, and cone shaped and are found on the undersides of leaves.

Silverleaf whitefly *(Bemisia tabaci)*

What kinds of plants do they attack?

Whiteflies will feed on nearly any plant but are especially pests on geraniums, greenhouse foliage plants, nicotianas, poinsettias, and other flowers as well as citrus, cotton, sweet potatoes, tomatoes, and other vegetable plants. Whitefly species vary in their preferred hosts.

Where are they found on the plant?

Whiteflies are found primarily on the undersides of leaves.

What do they do to the plant?

Whiteflies suck the sap from plants, leaving behind yellow, stippled damage, drooping leaves, and loss of vigor. They also vector a number of viruses and produce honeydew, a sugar-rich substance that promotes the growth of sooty mold and attracts ants.

General Discussion

Whiteflies are a pest in big numbers but are not difficult to get rid of. If you think the plants might have whiteflies but are unsure, shake the plants a bit. You'll see them flying off, then settling right back onto the leaves.

Females each lay about 100 tiny eggs on the undersides of leaves. Eggs hatch in about seven to 10 days, and the larvae develop into mobile, scale-

Whitefly damage on cucumber leaf; white or silver coloring usually covers the entire leaf.

like insects. They find a suitable feeding location and settle in, becoming nearly immobile and sucking sap from leaves. Larvae mature in two to four weeks and the adults live for four to six weeks after that. The reproductive rate is temperature dependent: most whitefly species do best in a temperature range of 80°F to 90°F (27°C to 33°C).

Exclusion and Prevention

➤ Inspect any new plants before introducing them into the garden or greenhouse.

➤ Prune and remove infested leaves and plants to curb an infestation.

➤ Use a HEPA filter in greenhouses.

➤ Keep greenhouse temperatures below 80°F.

➤ Use row covers.

CONTROLS

Azadirachtin	Neem oil
Capsaicin	Pyrethrum
D-limonene	Sesame oil
Garlic	Vacuuming
Herbal oils	Yellow sticky cards
Horticultural oil	(for indication purposes)
Insecticidal soap	

BENEFICIAL BIOLOGICALS

Beauveria bassiana

Big-eyed bugs

Damsel bugs

Lacewings

Lady beetles

Minute pirate bugs

Wasps, parasitoid *(Encarsia formosa)*

Interesting Facts

Whitefly populations rage out-of-control most often when the natural balance of prey and predator has been disturbed. Avoid spraying plants with broad-spectrum insecticides, as these kill beneficial insects as well as pests.

Whiteflies

2. DISEASES

Disease can strike cultivated plants at any stage. Most diseases that affect garden plants fall into two broad categories: fungal and bacterial. The spores and bacteria that cause plant diseases are ubiquitous. A garden's susceptibility to disease is often traceable to environmental imbalances in temperature, moisture, light conditions, airflow, pH, or nutrients.

Fungus grows when it finds the right moisture level, proper temperature (the range varies by species), acidic conditions, and a reliable source of food.

Bacteria are much more likely to invade when the environment has been compromised. Oxygen deprivation and wounds make their attack more likely to be successful.

Once disease hits, it is important to act quickly and restore balance to the environment. However, providing a balanced environment for the plants is the best solution.

There are several things that you can do to prevent your plants from being compromised by disease.

Seeds, seedlings, and plants: Soak seeds in compost tea, hydrogen peroxide, Spinosad, potassium bicarbonate, or colloidal silver to promote disease-free germination. These seed soaks can be used during the first period of growth. Spray plants every seven to 10 days when there is plant stress such as cool or warm, moist conditions.

At the end of each chapter is a summary of solutions specific to the disease. No store can carry all brands so familiarize yourself with the active ingredients listed. For more information on when and how to use the controls look them up in Section 5, which also includes referrals to commercially available products.

For many problems redundancy—using more than one solution—is good because it creates multiple modes of action.

INDOOR AIR FILTRATION AND SANITATION

Fungal spores often enter indoor gardens on air currents. A fine dust filter in the air intake system captures these spores and reduces the chance of fungal infections. Another option is a UVC lamp in the intake duct. The light from these lamps kills microbes and destroys spores. The two options can be combined.

Black Spot

How common is this disease?

Black spot fungus *Diplocarpon rosae* is the bane of rose growers in humid, moist environments. It germinates in seven hours under moist/wet conditions at 75°F. Symptoms appear three to 10 days later. It stops growth at 85°F.

Origin

Spores from the fungus overwinter on fallen, infected leaves and diseased canes. Spores are then spread to newly emerged foliage in spring by rains or watering.

Where is the disease found?

Leaves and stems, or canes, are affected.

Appearance and Effect of the Disease

Black spot appears on leaves as circular black spots with indistinct margins. The tissue surrounding the black spots yellows, possibly spreading to the whole leaf. On young canes, the fungus appears as purplish or black blisters. Plants drop affected leaves quickly, which slows the spread of the disease.

Black spot on rose

Healthy rose bush

What kinds of plants does it attack?

Black spot attacks only roses.

Exclusion and Prevention

Prevention is the key to controlling black spot.

➡ Plant resistant cultivars of roses. Hundreds of varieties have resistance, so choose from these.

➡ Keep the foliage dry. Water early in the day so that any water that gets on the leaves dries quickly. Prune plants and space new plantings to allow for maximum airflow.

➡ Remove affected leaves and canes. Do not compost them. Throw them away. Remove all plant debris at the end of the season, including fallen leaves and dead canes.

➡ If a black spot infection is expected, spray plants weekly with fungicidal soap, herbal fungicide, milk, or potassium bicarbonate, before symptoms emerge.

Typical infection

CONTROLS

Compost tea

Garlic

Herbal oils
(clove, coriander, and oregano)

Milk

Neem oil

Potassium bicarbonate: A 0.5 percent solution of potassium bicarbonate or sodium bicarbonate (1 teaspoon in 1 quart of water with a wetting agent helps control it.

Sodium bicarbonate

Sulfur

BENEFICIAL BIOLOGICALS

Bacillus subtilis

Streptomyces lydicus
(beneficial bacteria)

Damping Off (Stem Rot)

Damping off is a condition rather than a specific disease and is caused by several soil-borne fungi, including *Botrytis*, *Fusarium*, *Macrophomina*, *Pythium*, *Rhizoctonia*, *Phytophthora*, *Sclerotinia*, and *Thielaviopsis*. It is generally a disease of seeds and young plants and can occur in warm or cool soils, but it usually requires high levels of moisture.

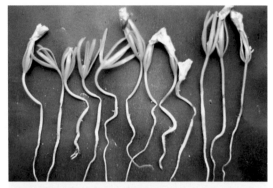

Stem rot

How common is this disease?

Damping off is a common problem in both indoor and outdoor growing environments and is widely distributed all over the world.

Origin

Damping off occurs in warm, nitrogen-rich soils that stay wet for long periods. Hydroponic systems can be sources of damping off when the growing media becomes too saturated. Damping off usually occurs when plants are placed in media already infected with one or more of the causative fungi.

Damping off fungi enter the seedling without the need for injuries or wounds. Once inside, the fungus releases enzymes to dissolve the cells. It then absorbs the nutrients released from the cells and grows deeper and deeper into the plant until the plant collapses and dies.

Where is the disease found?

Preemergent damping off affects seeds. They rot underground and never germinate. Postemergent damping off occurs after the seeds have sprouted. The infection begins at or below the soil line. Depending on the disease organism causing the problem, leaves, roots, or stems can be affected.

Appearance and Effect of the Disease

As damping off is caused by a number of fungi, symptoms vary. Generally, either seeds don't germinate and rot away in the soil, or young seedlings fall over and die. In seedlings, damping off is first apparent as a red, yellow, or brownish discoloration of the stem just above the soil line. Lesions and cankers form on the stem between nodes. As the disease progresses, the lower portion of the stem becomes mushy, soft, and brown. Eventually, it can no longer support the upper weight of the plant and it topples.

Stem rot

The wilting may resemble root rot when the wilt first starts and the leaves turn yellow, but stem canker has not yet appeared. The key difference is that damping off doesn't affect the roots.

In the later stages, the leaves droop and turn yellow. As lesions extend up the plant, it begins to wilt due to vascular damage. (See the entry on *Pythium* for more information on this.)

Damping off is sometimes mistaken for nutrient burn from overfertilization, salty soil, excessive heat, or cold or dry soil. Key diagnostic signs are the lesions and cankers on the stem, the brown discoloration, and soft stem tissue.

What kinds of plants does it attack?

Most types of fungi that cause damping off affect primarily seeds and seedlings. All species of plants are susceptible under the right conditions.

Exclusion and Prevention

➤ For greenhouse and indoor growing, always start with a sterile planting mix or pasteurized soil. Planting mixes and planting cubes are infection free. Only use sterile tools, pots, or trays for seedling production.

➤ Prevention is the main weapon against damping off.

➤ Avoid overwatering. Water only when the soil surface dries. Use a moisture meter or a finger to test for soil moisture. Testing near the edge of the container is less likely to disturb root development.

➤ Use soil mixes with vermiculite, perlite, or sand incorporated to provide adequate drainage.

- Nitrogen-rich soils hinder root growth and aggravate damping off.

- Plant seeds as shallow as possible as soil is wetter deeper into the pot, and with shallow planting it takes the seedling less energy to reach the surface.

- Provide adequate air circulation between plants to keep the environment as dry as possible to discourage damping off fungi.

- Encourage quick seedling growth indoors by providing adequate heat and light. Bottom heat hastens growth, making plants less susceptible.

- Apply a fungicide to the seeds before planting.

- Don't transplant seedlings outside until they have several sets of leaves. Younger plants have less resistance to pathogens.

- Dispose of any damping off–affected plant material as soon as possible.

- Use seeds that have been pretreated with fungicides.

- Avoid overfertilization, as nitrogen-rich environments encourage fungi growth.

- Incorporate properly aged compost and compost tea into soils for protection against damping off. Soak seeds in compost tea for a day before planting.

- Avoid watering seedlings by overhead means. Instead, place seedling flats into shallow pans of water and remove them when the top of the soil is damp. Use a fine layer of perlite or sphagnum moss on the stop of the soil to keep seedling stems dry at the soil line.

CONTROLS

Young seedlings affected with damping off stand little chance of recovery. Older seedlings with one or two sets of leaves stand a better chance of survival with the help of **fungicides**.	Kelp
	Milk
	Potassium bicarbonate
	Quaternary amines
Chamomile tea	Sesame oil
Compost tea	Sodium bicarbonate
Copper	UVC light
Herbal oils (clove, coriander, and oregano)	

BENEFICIAL BIOLOGICALS

Bacillus subtilis (beneficial bacteria)	*Mycorrhizae*
Gliocladium (beneficial fungus)	*Streptomyces griseviridis* (beneficial bacteria)
Pseudomonas (beneficial bacteria)	
	Trichoderma

Gray Mold and Brown Mold
(Botrytis cinerea)

How common is this disease?

Gray mold is found nearly everywhere and can cause disease on most plants. It is a cool-season disease and occurs both in greenhouses and outside.

Appearance and Effect of the Disease

A *Botrytis* infection first appears as a water-soaked, browned area no matter what type of plant tissue is affected. Soon, a silvery gray, fuzzy mat develops on the tissue, formed from thousands of grape-like clusters of spores. These spores fly up like dust when the affected area is disturbed. Leaves and buds yellow from being suffocated by the mold. In high humidity, leaves develop a brown slimy substance from the destroyed tissue.

What kinds of plants does it attack?

More than 50 species of *Botrytis* affect a wide variety of ornamental plants, including begonia, chrysanthemum, dogwood, fuchsia, hydrangea, marigold, pansy, periwinkle, rose, snapdragon, sunflower, violet, and zinnia. As for fruits and vegetables, it affects asparagus, beans, carrots, celery, cucurbits, eggplants, grapes, lettuce, peppers, raspberries, strawberries, tomatoes, and many others.

Botrytis on peony leaves

Botrytis cinerea sporulation on a ripe strawberry.

Where is it found on the plant?

Gray mold mainly affects tender tissues such as buds, flowers, and seedlings, but it can also enter the plant through wounds such as pruning scars or tissue weakened from age. The parts affected may depend on the species of *Botrytis* present.

Shaded areas of the plant that do not get a lot of light are usually first infected. Then the disease spreads quickly through the growth and spores.

Gray mold attacks many fall flowering plants, destroying their flowers when moisture from dew drops, rain, or watering is held in complex flowers.

What does it do to the plant?

Generally, *Botrytis* causes blights of blossoms, leaves, and seedlings; bud rot; stem canker; stem and crown rot; and damping off.

General Discussion

Gray mold is an environmental disease. As long as the humidity stays at 50 percent or lower, gray mold is unlikely to occur. The fungus can germinate only on wet plant tissue when the temperature is between 55°F and 70°F. This can happen in dry weather as dew covers the leaves. Once it starts growing, the fungus tolerates a wide range of humidity and temperatures, but high humidity and cool temperatures help it thrive. Lowering the humidity often stops its growth. Gray mold feeds on the plant tissue, collapsing the cells and causing the tissue to rot. Some species of *Botrytis* form sclerotia, the overwintering stage of the fungus. In late summer, these tiny black structures may be visible on affected plants.

A strawberry rachis completely engulfed by a gray mold fungus, *Botrytis cinerea*.

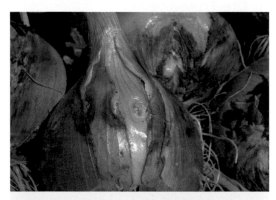

Onion bulbs with symptoms of the *Botrytis* soil line rot.

Botrytis blight of Celosia

Like most other fungi, gray mold enters plant parts wounded or damaged by insects or pruning, or plants that are dying. For this reason, it is important to sanitize pruning equipment between cuts.

Cuts and lesions are a normal part of plant life, so all plants are subject to attack when conditions are favorable to the mold.

Spores travel mostly via wind, rain, and drip from infected vegetation. Indoors and out the spores are endemic in the air and can also be carried by clothing and pets.

Exclusion and Prevention

- Avoid favorable conditions by keeping humidity under 50 percent and temperatures higher than 75°F in indoor growing areas such as greenhouses.

- Provide adequate air circulation both indoors and out through pruning and planting distance to help control the humidity and temperature.

- Remove and discard dead, dying, or diseased plant tissue.

- Avoid overwatering both indoors and out, especially when conditions are favorable for germination.

- Avoid splashing water on foliage when watering. Water at soil level and early in the day so that any splashed water has a chance to evaporate.

- Use sterile pots and unused or pasteurized planting mix for indoor growing.

CONTROLS

Apply any of these controls preventatively whenever plants are threatened with attack. All of these fungicides can also be used postemergence to stop the spread of disease. They may be washed off by rain so they should be reapplied after each storm.

Chamomile tea	Neem oil
Compost tea	ph Up
Copper	Potassium bicarbonate
ELISA test kit	Quaternary amines
Herbal oils (clove, coriander, oregano, and thyme)	Sesame oil
	UVC light
Milk	

BENEFICIAL BIOLOGICALS

Bacillus subtilis	*Streptomyces lydicus* (beneficial bacteria)
Clonostachys rosea	*Trichoderma harzianum*

Powdery Mildew

How common is this disease?

Powdery mildew is a fungal disease that affects a wide range of plant species. Each species of powdery mildew has a limited host range, but all are characterized by an easily recognizable white or gray powdery growth.

Mildew spores can be found everywhere. It is a common problem in both indoor and outdoor gardens whenever the temperature and humidity fall into its favored range.

Powdery Mildew on Squash Leaves

Origin

Powdery mildew spores are airborne, are found everywhere, and can lie dormant until the environmental conditions are right. Initial powdery mildew infections are caused usually by spores transported to new plants by the wind. Mycelia and hardened spores overwinter in plant debris and infected plant parts. Pets, clothing, and plant introductions are also vectors.

The disease thrives in these conditions: a slightly acidic surface, moderate humidity to high humidity (but not wet conditions), temperatures of 60°F to 85°F (depending on species and race), and low light intensity.

Powdery Mildew on Sweet Pea Vine

Where is the disease found?

Powdery mildew primarily attacks new leaves but can also affect buds, fruit, stalks, stems, and growing tips. The infection spreads over the plant and to other plants.

Appearance and Effect of the Disease

Powdery mildew is one of the most recognizable plant diseases in the garden. At first it appears on just a small portion of the leaf in an irregular circle pattern. It quickly spreads and soon the entire leaf is covered as if it had been powdered with confectioner's sugar. Later the leaves become distorted and stunted, turn yellow and brown, and drop. The taking of nutrients from the plant by the fungus results in a general decline in plant health.

What kinds of plants does it attack?

The host range is wide. Plants particularly affected include begonia, dahlias, peas, phlox, roses, squash, and zinnias.

Powdery Mildew on Hydrangea

Powdery Mildew on Rose Leaves

Exclusion and Prevention

► In greenhouses, quarantine all new plants in a separate area where they can't infect other plants; use a germicidal UVC light like the ones used in food handling—it kills airborne spores.

► In greenhouses and outdoors, restrict garden humidity by watering early in the day.

► Space plants far enough apart to allow free airflow.

► Powdery mildew thrives in hot temperatures with cool nights. Indoors, keep nighttime and daytime temperatures close together.

► Drought-stressed plants are more susceptible to powdery mildew than those adequately watered.

► Remove plant debris from the garden as soon as it is created.

Powdery Mildew on Lilac

➤ Remove infected leaves carefully without knocking spores into the air. Place a bag over infected leaves and tie it shut; then cut off the branch. Use a fungicide on wounded branches.

➤ Fungi die in alkaline environments. Adjust the pH of the leaf surface to a pH over 8.

CONTROLS

Compost tea and compost

Copper

D-limonene

Fish oil

Garlic

Herbal oils (clove, coriander, oregano, and thyme)

Horticultural oils (jojoba or cottonseed oil)

Hydrogen peroxide

Milk

Neem oil

ph Up

Potassium bicarbonate

Quarantine new plants

Sesame oil

Sodium bicarbonate

Sulfur

UVC Light

Vinegar

BENEFICIAL BIOLOGICALS

Ampelomyces quisqualis (beneficial fungi)

Bacillus pumilus

Bacillus subtilis

Streptomyces lydicus (beneficial bacteria)

Fungicides with different modes of action can be combined to increase effectiveness. Here Serenade and Ed Rosenthal's Zero Tolerance Fungicide were used together.

Anti-fungal spectrum lamps from Clean Light control powdery mildew when used daily.

Root Diseases

Fusarium Wilt

Phytophthora

Pythium

Rhizoctonia

Verticillium Wilt

Every plant must have a healthy root system. Pathogens can attack and damage the roots of one plant, then rapidly infect other plants in the garden. *Fusarium, Phytophthora, Pythium, Rhizoctonia,* and *Verticillium* are common and destructive root aggressors known to all growers, whether soil gardeners or hydroponicists.

Roots and stems are most likely to be attacked by pathogens when they are stressed and weakened. The main cause of stress is usually a lack of oxygen, which the roots require. Soggy conditions and stagnant water are usually oxygen-deficient situations. The solution is to make sure that the soil is well drained, the stems have a chance to dry out, and water is oxygenated.

The beneficial fungi mycorrhizae help protect plant roots by forming associations that physically surround them, so pathogens cannot reach them. The fungi sometimes also attack pathogens. It is worked or watered into the soil, helping to protect the roots and making the soil's resources more available to the plant.

Compost tea contains enzymes and hormones as well as living organisms that help stimulate plants' immune systems and also directly protect roots from pathogens. When it is used regularly, plants are much less likely to fail.

Hydrogen peroxide is an alternative solution. When used as a 1 percent solution it kills pathogens. Colloidal silver can also be used to keep water pathogen free.

Vital Earth's Mycorrhizal forms symbiotic relationships with roots increasing their ability to absorb nutrients, and more importantly protects them from disease-causing pathogens.

Vital Earth's PHC BioPak contains nutrients and bacteria that create a healthy root environment while stimulating growth.

Root Diseases:
Fusarium Wilt

How common is this disease?

Fusarium wilt is a fungal, soilborne disease that is caused by the *Fusarium* genus. It is specific to certain crops. It is a common, or cosmopolitan, fungus that thrives in warm temperatures.

Fusarium wilt is not commonly found in gardens, but residual spores remain in soil where it has appeared. Since it is soil based and species specific, it may be common in one space but rare in another, very close space.

Fusarium wilt symptoms on sugar beet

Where is the disease found?

Fusarium wilt is found in the soil and infects plants through the rootlets, then travels up the plant by way of the xylem until it eventually affects the whole plant. It can also be seed borne.

Appearance and Effect of the Disease

Symptoms usually appear on the lower and outer leaves first as small, dark, irregular spots. Then leaves wilt and yellow, and their tips turn upwardly and hang off the stem but don't fall off. Frequently, the symptoms appear only on one side of the plant. Cut stems exhibit a brown discoloration in the interior. *Fusarium* species produce somewhat different diseases. These diseases have slightly different symptoms but respond to similar prevention and control methods.

Fusarium root rot begins below the soil line, turning the roots rotten and necrotic and giving them a characteristic red color. The first visible symptom usually appears as the rot works its way up the stem, producing a red-brown discoloration; it may progress to swelling and the stem may split open. The plant soon begins to wilt, then collapses as the decay spreads up the stalk.

In both wilt and root rot, the fungus spreads through plant cells and clogs the xylem vessels, inhibiting water and nutrient transport. This vascular clogging inside the plants causes the external symptoms of wilt and collapse. Infected plants usually die.

What kinds of plants does it attack?

Fusarium wilt affects a wide range of plants, many specific to certain *Fusarium* species. Some common hosts are dahlias, hops, melons, mimosa trees, peas, peppers, tomatoes, and watermelon.

Exclusion and Prevention

➤ Plant resistant varieties.

➤ *Fusarium* survives in plant debris, so it should not be buried, composted, or placed on uninfected soil.

➤ Sterilize trays and soil, especially if used for indoor growing where warm conditions will increase the incidence of infection.

➤ Clay soils have been found to be fungistatic due to their high pH. Also, loamy soils with diverse and healthy plant growth harbor beneficial microbes that help to suppress *Fusarium*. These soils do not stop the fungus, but they slow it down.

➤ Adding well-cured compost or compost tea to soil can help protect plants. Microbes in compost tea are antagonistic to *Fusarium* fungi and help reduce infections.

➤ Make sure the pH level doesn't get too

A common hop plant showing foliar symptoms of *Fusarium* canker

A common hop plant showing foliar symptoms of *Fusarium* canker

Fusarium wilt on mum root, showing vascular discoloration

low. Alkaline conditions discourage fungal growth. Neutralize acidic soils with dolomite lime or greensand.

➤ Fertilizers enhanced with potassium or calcium help prevent *Fusarium*. Excess nitrogen levels can encourage fungal growth.

➤ Rotate crops. Avoid planting the same crop in the same ground for many years in a row. Even if none of the plants show symptoms, multiple successive plantings can cause the fungus to build up in the soil until it reaches destructive levels.

➤ Mycorrhizae fight pathogens and help to improve plants' disease resistance.

➤ *Bacillus pumilus, Bacillus subtilis,* and *Streptomyces grisoviridis* (all beneficial fungi) and *Gliocladium* (a beneficial bacteria) can be applied as a pretreatment soil drench for seeds, as a soil treatment, or as a foliar spray.

Roots of dry beans showing advanced symptoms of *Fusarium* root rot

CONTROLS

➤ There are no true controls for *Fusarium* once it has invaded crops. The only effective measure is to destroy the affected plants and sterilize pots, tools, and any growing surfaces, such as bench tops.

Onion plants showing *Fusarium* wilt

BENEFICIAL BIOLOGICALS

➤ Beneficials such as *Streptomyces lydicus* (marketed as Actinovate) will help prevent infection and lessen the disease's effects if it's caught early.

Root Diseases:
Phytophthora

How common is this disease?

Phytophthora is a common disease in gardens, nurseries, and natural settings such as woods, parks, and forests.

Origin

Phytophthora is a group of diseases caused by oomycetes (water molds), a life-form with its own kingdom. Until recently, they were classified as fungi, but genetic research has shown that

Zucchini squash plant showing symptoms of *Phytophthora* wilt.

though they show parallel forms and styles, they are a fundamentally different life-form. One aspect of this is that while fungi build their cell walls using chitin, which is also used by insects for structure, oomycetes use cellulose, the same material plants use for structure.

Oomycetes are found in soils throughout the world, but they require moist even wet conditions in order to infect plants.

Appearance and Effect of the Disease

Depending on the species of oomycete, any part of a plant may be infected. The disease causes the collapse of cells. They turn brown and sometimes mushy. By the time symptoms of the disease are apparent, the plant may be in final stages of collapse because the root system, though infected, can still provide nutrients and water to the canopy. As the infection grows, the supply line collapses. Upper portions of the plant may become infected as well. It is typified by brown patches on the leaves. In trees, it causes dieback at the tips of the branches, which travels in toward the core of the plant. A typical example is sudden oak death caused by *Phytophthora ramorum*.

Phytophthora die back on rhododendrum

What kinds of plants does it attack?

Phytophthora attacks all dicots, which are plants that produce two cotyledons, seed leaves, when they germinate. This includes trees such as firs, fruit trees, oaks, and pines as well as many other garden plants such as crucifers (cabbage, broccoli, etc.) and nightshades such as pepper, potato, and tomato. *Phytophthora* infestants cause potato blight, which devastated Europe, especially Ireland from 1845 to 1852. It also causes late tomato and pepper blight. Cucurbits such as cucumber, pumpkins, and squash are affected by the cosmopolitan *Phytophthora capsici.* It affects many plant groups, causing foliar blight, fruit rot, leaf spots, root and crown rot, seedling damping off, and stem lesions. Leaf spots are dark brown and up to 2 inches (5 cm) in diameter. Stems appear mushy or water soaked, are shades of brown, and are collapsed.

Many other garden plants, both ornamentals and fruit and vegetables, are affected by *Phytophthora* species, including oaks, which have been devastated by sudden oak death caused by *Phytophthora ramorum.*

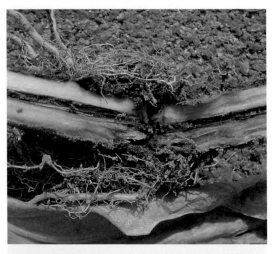

Pepper plant infected with *Phytophthora*

Where is it found on the plant?

The blight can be found anywhere on the plant depending on the host, the species attacking it, and where on the plant the infection occurred. Plants usually are infected in one of two ways: through irrigation water or by infected water splashed on the leaves or stem. However, some types of *Phytophthora* spores are airborne and can land on a leaf in a drop of rainwater. *Phytophthora* also produces "zoospores," another type of reproductive cell that swims and is attracted to plant roots.

Cabbage stem split to see decay of pith tissue of the lower stem

What does it do to the plant?

Wherever the infection gets started it begins its path of destruction. For instance, it might infect a fine root thread. As it destroys the tissue it follows the thread to a larger root, then travels up the stem and eventually collapses the canopy. Only some parts of the plant may be affected at first, because the infection has not traveled yet to the whole plant. As the organism grows, it leaves death in its wake, eventually killing the plant.

Exclusion and Prevention

➤ The main way to stop the spread of this disease in a growing area is to make sure that the irrigation/water system doesn't promote the transfer of the infection.

➤ Make sure the water is free of infection. Have it tested or use an ELISA kit.

➤ Remove all infected plants.

➤ Don't leave container plants in standing water, and prevent standing water in fields.

➤ In orchards keep water away from the tree trunks and leave a bare dry area of 18 inches around the trunk.

➤ Avoid spraying leaves when fertilizing. If you do spray leaves, do it early in the day so they dry thoroughly.

➤ Recycled water should be treated to kill pathogens. Use chlorine, hydrogen peroxide, ozone treatment, quaternary amines, or UVC sterilization.

CONTROLS

➤ There are no satisfactory chemical controls.

➤ Treatment with phosphite, a fertilizer absorbed by plants, seems to afford some protection to plants that have not been affected. Some researchers found that plants made a partial recovery with its use and that other plants exposed to infection remained healthy.

➤ Use varieties of plants that have shown resistance to the disease.

BENEFICIAL BIOLOGICALS

➤ Beneficials such as *Streptomyces lydicus* (marketed as Actinovate) will help prevent infection and will help lessen the disease's effects if it's caught early.

Root Diseases:
Pythium

Pythium is a destructive parasitic root fungus. Under favorable conditions, it multiplies very rapidly, releasing microscopic spores that infect the roots and depriving the plant of food. It attacks mainly seeds and seedlings, which have little resistance. It stunts older plants that it infects.

How common is this disease?

Pythium is a common problem in field, container, and hydroponic cultivation.

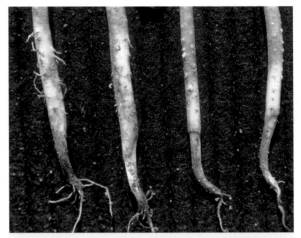

Dry bean seedlings showing symptoms of *Pythium* damping off caused by *Pythium spp.*

Origin

Pythium is resident in most soils no matter how clean your garden. It is a "secondary infection"; it attacks plants that are already stressed due to disease, damage, pests, nutritional deficiencies, or poor growing environments.

Pythium develops most efficiently with high humidity and temperatures between 70°F and 85°F. Lack of oxygen in hydroponic nutrient solutions contributes to *Pythium* development.

Where is the disease found?

Pythium generally affects the entire plant but is mainly a root disease. Seedlings and young plants are most vulnerable.

Appearance and Effect of the Disease

Affected plants wilt, yellow, or appear stunted, or the leaf edges are sometimes browned.

The roots become discolored, soft, and watery, depriving the plant of food. As the infection advances, the outer tissue of the root sloughs off, exposing the stringy, inner core. A *Pythium* infestation can advance to the crown of the plant and cause the same discoloration and softening that it produces in the roots.

Pythium moves through the soil or water to the plant roots where it germinates before entering the roots. Once it invades, it spreads through the tissue and produces spores that germinate inside the plant. This causes severe wilting.

What kinds of plants does it attack?

Pythium mainly attacks seeds, seedlings, or juvenile growth.

Exclusion and Prevention

➡ Keep plants healthy and free of other pests to avoid giving *Pythium* a chance to take hold. Weakened plants are prime candidates for attack.

➡ Sterilize pots and tools before planting and after use.

➡ Avoid overwatering and overfertilization.

➡ Use well-drained pasteurized soils and soil mixes.

➡ Keep fungus gnats in check in indoor growing areas as they can carry *Pythium* into a sterile area.

➡ In hydroponic systems, prevention of *Pythium* is particularly important as *Pythium* can spread quickly once introduced. Keep water clean to avoid infection. Treat the water with hydrogen peroxide (H_2O_2) or use UVC light to kill waterborne spores. The down side of these treatments is that both also kill beneficial organisms.

CONTROLS

Compost tea	Herbal oils (clove, coriander, oregano, and thyme)
Copper	Quaternary amines
ELISA kit	UVC light

BENEFICIAL BIOLOGICALS

Preventative biocontrol agents to apply before plants show symptoms include:

Bacillus subtilis (beneficial bacteria)	(beneficial bacteria)
Gliocladium (beneficial fungus)	*Streptomyces lydicus* (beneficial bacteria)
Pseudomonas (beneficial bacteria)	
Streptomyces griseoviridis	*Trichoderma* (beneficial fungi)

Root Diseases:
Rhizoctonia

How common is this disease?

It is a common soilborne disease throughout the world.

Origin

Rhizoctonia is the asexual form of a fungus.

Appearance and Effect of the Disease

It is one of the causes of damping off, which begins at the root, where the fungus invades the root. The fungal hyphae invade living cells that they feed on and also produce digestive juices that destroy surrounding cells so their nutrients can be used by the fungus.

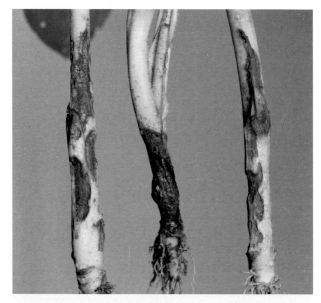

Root rot/damping off. *Rhizoctonia spp.* Symptons are sores on outer stem that spread inward

The fungus moves up the plant and attacks the stem, causing a rot that appears as a darkened color. At the same time, the stalk begins to rot, giving it the name "sore shin." As the fungus continues to grow, leaves yellow and wilt, and eventually, the plant topples over from a weakened stem.

What kinds of plants does it attack?

It attacks a wide range of plants, usually in the emergent or seedling stage. Common plants affected include cereals, flowers, grains, and vegetables.

Where is it found on the plant?

It usually invades plants at the root level and then grows upward to the stem.

What does it do to the plant?

It destroys and digests plant cells, beginning with the roots and climbing up the stem. It continues growing upward but soon kills the plant.

General Discussion

Rhizoctonia solani is one of the most widespread pathogens of plants worldwide. Plants are most susceptible when they are first starting out, but this pathogen may also pose a danger when plant roots are disturbed, such as during transplanting if the soil is infected.

Exclusion and Prevention

➤ Use sterile or pasteurized planting media to start seeds. Long-lived *Rhizoctonia* spores can survive in soils for more than five years.

➤ Plant seeds and transplants in warm weather, so they grow quickly and are not stressed.

➤ Use a soil high in calcium.

➤ Compost and compost tea thwart *Rhizoctonia*. Add compost to the mix and water with compost tea.

➤ Brassicas such as cabbage and broccoli thwart *Rhizoctonia* because they release iso-thioeyanates, which are fungicidal. Plant them after your crop has been harvested and turn them over before frost.

CONTROLS

Compost and compost tea

ELISA kit

Herbal oils (clove, coriander, oregano, and thyme)

UVC light

Wood ashes

BENEFICIAL BIOLOGICALS

Bacillus pumilis

Bacillus subtilis

Pseudomonas (beneficial bacteria)

Streptomyces griseoviridis (beneficial bacteria)

Streptomyces lydicus (beneficial bacteria)

Trichodermas such as *Gliocladium virens* and *harzianum*

Rhizoctonia lesion on the lower stem of a pepper stem

Root Diseases:
Verticillium Wilt

How common is this disease?

Verticillium wilt is caused by soilborne fungi that are found in all soils.

Origin

Verticillium wilt thrives in moist, poorly drained soils such as clays and is more common in cool weather.

Where is the disease found?

Infection begins through attacks on stressed roots and travels up the plant to the leaves.

Appearance and Effect of the Disease

The lower leaves turn yellow along the margins and between the veins. The yellowing sometimes extends to the whole leaf. Then they turn gray brown, wilt, and fall off. As the disease spreads upward, it clogs the vascular system, reducing the water flow through the plant. The stem turns brown, and the wilt spreads up the plant. Leaves fall off early, and the plants die. A cross section of the lower stem near the crown of the plant reveals a discolored area. Symptoms may resemble *Fusarium* wilt.

Common hop plants showing foliar symptoms of *Verticillium* wilt

Healthy looking strawberry plant (left); the other is infested with *Verticillium* wilt.

What kinds of plants does it attack?

Verticillium wilt attacks a wide variety of woody and herbaceous plants, including asters, cherries, chrysanthemum, hops, maples, melons, peaches, pepper, strawberries, and tomatoes.

Verticillium wilt is one of the most common and destructive diseases of shade and ornamental trees and bushes.

Sunflower plants showing symptoms of *Verticillium* wilt

Exclusion and Prevention

➡ Plant varieties resistant to *Verticillium* wilt. Look for plant marketing tags marked with the initials VFN. V stands for resistance to *Verticillium* wilt, F stands for resistance to *Fusarium* wilt, and N stands for resistance to root knot nematodes.

➡ As tomatoes are particularly affected by *Verticillium* wilt, seek resistant cultivars, including Beefmaster, Better Boy, Big Girl, and Rutgers 39, among others.

➡ Remove and destroy all infected residues.

➡ Use sterile planting mix and sterilize growing systems.

➡ Amend soil with alfalfa meal or aged compost.

➡ Add mychorrizae and trichoderma to the planting mix; they create a barrier to pathogens around the roots.

➡ Avoid overwatering and use well-drained planting mix.

➡ Add aged compost to the soil mix and water with compost tea.

➡ Use fertilizer low in nitrogen and high in potassium.

➡ Do not plant susceptible plants in areas where *Verticillium*-infected plants were growing.

CONTROLS

There are no chemical controls available.

BENEFICIAL BIOLOGICALS

Bacillus subtilis (EZB24 strain)	*Trichoderma* (beneficial fungus)
Streptomyces lydicus (beneficial bacteria)	Mycorrhizae

Septoria Leaf Spot

How common is this disease?

Septoria leaf spot is a fungal disease that occasionally attacks garden crops, especially tomatoes, but is far more serious on farms. *Septoria* is specific to the host. The species that attacks tomatoes doesn't attack cucumbers or lettuce. It is not usually a problem indoors.

Origin

Septoria attacks in wet, warm, cloudy weather. Warm spring and summer rains trigger the release of spores from the overwintering storage structures. Infection occurs when temperatures are in the 60s °F, but the fungus grows faster and becomes more destructive as the temperature rises, the ideal being just below 80°F.

Septoria leaf spot on tomato leaf

Septoria leaf spot and powdery mildew infecting celery leaf

Where is the disease found?

Septoria attacks the lower leaves of garden vegetables, such as tomato and others, first. In landscape plants and trees, the spots first appear scattered throughout the foliage. Spotting may also occur on the blossoms and the stem.

Appearance and Effect of the Disease

The spots may first appear as yellow water-soaked spots that eventually turn brown. Depending on the species of plant, they may also first appear as tan, white, or gray. Spots may also have a halo of yellow or brown tissue and often have a pimple of black tissue in the center. These spots enlarge to a width of 1/16 to ¼ inch in diameter and often move up the stem. Whole leaves may turn yellow and fall off. Plants lose vigor with the loss of leaves. In fruiting plants, such as tomato, the reduction in leaf cover can result in sunscald to the fruit.

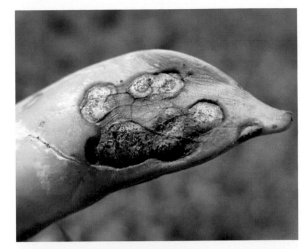

Pod lesion of *Septoria* blight on dry pea

What kinds of plants does it attack?

Septoria attacks a very wide range of plants, including dogwood, pistachio, potato, rhododendron, *Rudbeckia*, shade trees, small grains, tomato (where it causes severe blights), and wheat.

Exclusion and Prevention

➤ Remove plant debris to eliminate overwintering sites.

➤ Use disease-free seed.

➤ Rotate crops with nonhost plants.

➤ Control weeds in and around the garden to eliminate alternate hosts.

➤ Thoroughly till the garden after harvest to break up any remaining infected debris.

CONTROLS

Compost and compost tea	Potassium bicarbonate
Copper	Sesame and fish oils
Milk	Sodium bicarbonate
Neem oil	Sulfur
ph Up	

BENEFICIAL BIOLOGICALS

Bacillus pumilus (beneficial bacteria)	Bordeaux mixture
Bacillus subtilis	*Trichoderma* (beneficial fungi)

Tobacco Mosaic Virus

How common is this disease?

Tobacco mosaic virus (TMV) occurs occasionally in gardens, greenhouses, and indoors. Its two vectors are contamination by handlers and transfer by pests.

Origin

TMV is the first virus that was described scientifically. It is common on tobacco that is sold commercially such as in cigarettes. The virus remains viable after the tobacco is cured, dried, treated with chemicals, and even if frozen. Handling tobacco or even just smoking a cigarette becomes a vector for the disease. Even if one's hands are washed thoroughly after exposure, clothing remains a possible source of casual exposure.

Tobacco Mosaic Virus on tomato leaf

Discarded cigarette butts can also transfer the virus.

Once it has made an appearance in a garden, it can be transferred to other plants through touch, handlers, and pests.

Appearance and Effect of the Disease

Leaves affected by TMV develop a mosaic of darker and lighter green areas or yellow shades. In some plant varieties, leaves become thinner. Fruits and flowers may be scarred.

The disease is more likely to affect the vitality of the plant, rather than kill it.

What kinds of plants does it attack?

Common plant hosts for the mosaic virus are the nightshade and cucurbit families. Plants include delphinium, marigold, pepper, petunia, snapdragon, and tomato. It also infects celosia, cucumber, ground cherry, impatiens, ivy, jimson weed, muskmelon, nightshade, phlox, plantain, spinach, squash, and zinnia.

Where is it found on the plant?

The newest plant foliage shows symptoms first, but the whole plant becomes infected.

What does it do to the plant?

On tomatoes the leaves become bumpy and mottled with light and dark green patches and some yellowing. The leaves develop blister-like lesions. The leaves grow thin and distorted, with a resemblance to fern leaves. The edges of the plant leaves crinkle, and the blossoms become discolored. Fruit production will be stunted, and the few tomatoes produced by the diseased plant will be misshapen and blemished.

Similar affects occur on other plants.

General Discussion

TMV can devastate a garden or farm. Since contamination and spread of the disease can come from casual contact, it is important to develop a protocol to prevent accidental exposure by humans, pets, infected plants, or refuse.

Tobacco Mosaic Virus on petunia leaf

Tobacco Mosaic Virus on Euphorbia

Exclusion and Prevention

➤ A periodic, thorough inspection of all plants in the garden for TMV is the only way to ensure that it remains virus free. An ELISA kit can be used to determine if the virus is present. Any suspect plant and its neighbors should be removed. After this, thoroughly wash and change cloths because a plant's casual brush with the virus is all that is needed to infect it.

➡ TMV is a disease that has been spread by human interaction and cultivation. It is preventable by using a strict no-tobacco policy before or while working in the garden and by making sure that new introductions are free of TMV infection.

CONTROLS

➡ There are no cures for infected plants. Prevention by removal is the only course of action.

➡ Remove sick plants without delay before they contaminate others.

3. NUTRIENTS

Nutrient disorders can occur in any growing medium while using any technique—planting in the ground or in containers with planting mix or growing hydroponically. Outdoor growers usually have fewer issues with nutrient disorders than indoor growers, but a lack of nutrients or a nutrient imbalance slows growth, limits flower and fruit production, and detracts from the plant's or garden's vigor and appearance.

Nutrient problems can be caused by a variety of issues: pH is one huge factor. Nutrient deficiencies make the plants sick. An overabundance of nutrients can result in nutrient burn or toxicity, and it can also lock out other ingredients.

All fertilizer packages list three numbers that identify the N-P-K ratio. They usually appear as three numbers with dashes between them such as 25-10-10. The first number represents nitrogen (N), which is responsible for foliage or leaf development. Fertilizers that promote heavy leaf growth have a higher first number than the other two. The second number is phosphorus (P), which is important for strong stems and flowering. The third number is potassium (K), which promotes healthy metabolic function. Sometimes micronutrients are listed after the macronutrients: these are calcium (Ca), copper (Cu), iron (Fe), magnesium (Mg), manganese (Mn), and zinc (Zn). Fertilizer mixes can contain a variety of helpful ingredients. For instance, MycoMinerals Organic Soil Amendment and other brands contain N-P-K, a variety of trace minerals, and additionally, mycorrhizae, a root inoculant that supports the vitality of the roots, and boosts the roots' ability to take in nutrients.

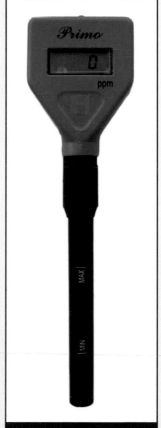

EC meters measure the parts per million (PPM) of nutrients in irrigation water or hydroponic solutions. They help determine how much fertilizer is needed. Hanna's Primo 2 is easy to use because it has one button calibration.

The ability to absorb nutrients is related to the pH of the water given to the plant. pH is the measurement of how acidic or how basic (alkaline) something is. pH should be viewed like a seesaw. It is important to measure pH after adding nutrients. pH can be adjusted by using commercially available pH-up or pH-down mixtures. Home remedies are available but can cause problems. Commercial products tend to be more concentrated and quite inexpensive. Failing to manage pH levels can cause nutrients to be unavailable and basically wasted. pH is important for soil gardening and so important for hydroponic gardening that failing to monitor pH has disastrous results.

In this section, we list the nutrient deficiencies your plants are most likely to encounter, in alphabetical order.

We start each nutrient description with a quick reality check: How common is this problem? This is followed by a description of how the plant appears if affected by that particular deficiency. Next is information

MycoMinerals Organic Soil Amendment contains N-P-K, trace minerals, and mycorrhizae, which aid the roots in nutrient uptake while protecting them from pathogens.

on its mobility and the role that the nutrient plays in the plant's nutrition. By mobility we mean that once a nutrient is transported to a site and is used to build tissue, can it be moved or is it fixed? Mobile nutrients move to new growth so deficiencies appear on older vegetation. Nonmobile nutrients stay put so that deficiencies appear on new growth. Each nutrient concludes with a guide to fixing the deficiency and getting the plant back to full health.

Unless the damage is slight, individual leaves do not recover from nutrient deficiencies. With nonmobile nutrients, look at new growth to determine whether the deficiency has been solved. With mobile nutrients, especially nitrogen, the damage will be seen first in the older leaves and not in new growth.

pH

pH is the logarithmic measure of the acid-alkaline balance in soil or water. A pH of 0 is the most acidic or "sour." Seven is neutral, and 14 is the most alkaline (basic or "sweet"). Generally speaking, the most desirable pH for good garden soil is 5.8 to 6.5. In this range, all of the essential plant nutrients are water soluble, making them readily available to plants. See pH entry in this section for more information on using pH to determine and ensure your garden's health.

Boron (B)

Early sign of deficiency in tomato leaf from top of plant. Notice the curl as well as the yellowing. Next the leaf will wilt, then die.

How common is this deficiency?

Boron deficiency occurs most frequently in some western soils. It may occur under a wide range of soil conditions, including high pH, in acidic and sandy soils that contain low levels of organic matter and excessive rainfall, and in dry soils, such as in drought conditions.

Symptoms

Boron-deficiency symptoms may appear as a shortening of the terminal growth, discoloration of the leaves, bushy new growth, failure to flower, then wilt and death. In root crops, such as beets or turnips, the interior tissue is dark and corky.

Common sign of deficiency in apples and other fruits and vegetables: Rotten brown circles appear on the skin and soft brown spots develop inside the fruit.

Mobility

Boron mobility varies among plant species, but in most plants, it is poorly transported.

Role in Plant Nutrition

The primary role of boron is the regulation of carbohydrate metabolism in plants. It is also essential for the formation of cell walls and seed development, germination of pollen grains, formation of pollen tubes, sugar transport, and flower retention.

CORRECTION

➤ Boron deficiency may be corrected with a foliar application, or with an addition to irrigation water. The recommended treatment rate is 1 teaspoon of boric acid (available in drugstores) per gallon of water. Borax, compost, and compost tea are fast-acting solutions.

Calcium (Ca)

Blossom end rot of tomatoes, squash, and other fruit is a common deficiency symptom.

How common is this deficiency?

Outdoors, calcium deficiency is rare except in very acidic soils. A calcium deficiency may not always be caused by a lack of it in the growing medium. Calcium is dissolved in water moving through the plant and is translocated from the roots to the leaves. Inadequate water or a lack of water movement due to cool, cloudy weather can cause a deficit in the leaves and fruit and problems such as blossom end rot in squashes and tomatoes. Excess nitrogen can also contribute to calcium deficiency by causing the rapid growth of plants, which draws extra water and nutrients to the leaves, starving other parts of the plant (such as the fruit) of calcium. Heavy rainfall may also wash calcium out of the root zone.

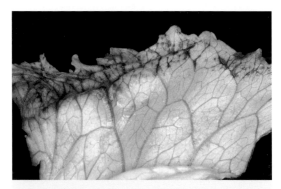

The new growth of lettuce exhibits necrosis that spreads to older leaves.

In greenhouses, calcium deficiency sometimes occurs in soilless growing mediums that have not been supplemented with lime, which is composed mostly of calcium. Distilled water and reverse-osmosis water, as well as some tap water, lack significant amounts of dissolved calcium. This can lead to calcium deficiency unless the water is supplemented with calcium.

Symptoms

Calcium-deficiency symptoms are most often first observed in the rapidly growing part of the plant such as the stem, leaf,

Yellowing of the edges of hop leaves and many other plants is the first symptom of deficiency. This spreads as the parts affected turn brown.

and root tip. Calcium deficiency stunts growth. Leaves turn dark green. Large necrotic (dead) blotches of tan, dried tissue appear mostly on new growth but also on other plant parts along leaf edges. Young shoots crinkle and get a yellow or purple color. In severe cases, they twist before they die. Necrosis appears along the lateral leaf margins. Problems migrate to the older growth, which browns and dies. Stems and branches become weak, lack flexibility, and crack easily. The root system does not develop properly, leading to bacterial problems that cause root disease and die-off. The roots discolor to a sickly brown. A characteristic hooking at the end of the leaf is an easily recognizable symptom. Roots

Nutri-Cal is derived from plant extracts, contains no chlorides, and has a low salt index making it safer to use. For faster action use as a foliar spray.

become short and stubby and turn brown. The most famous calcium-deficiency affliction is blossom end rot of tomatoes and other fruit. Plants chronically deficient in calcium have a tendency to wilt easier than plants not deficient.

Mobility
Calcium is semi-mobile.

Role in Plant Nutrition
Calcium strengthens plant cell walls and therefore stems, stalks, and branches, and it aids in root growth—mostly the newer root hairs. It travels slowly and tends to concentrate in roots and older growth. Calcium also enhances the uptake of potassium.

CORRECTION

► Add dolomitic lime, or garden lime, to potting mixes before potting plants. This provides calcium and can help stabilize the pH of acidic soils within a range of 5.9 to 6.5. One teaspoon of hydrated lime per gallon of water provides relatively fast absorption. Dolomitic limestone, which contains magnesium and calcium, takes longer to absorb. It is a good ingredient to place in planting mixes to prevent deficiency. Gypsum, calcium sulfate ($CaSO_4$), can be added to the soil to increase calcium content without affecting the pH too much. It should not be added to soils with a pH below 5.5 because it interacts with aluminum, making the metal soluble and poisonous to the plants.

► Calcium nitrate ($CaNO_3$) is a water-soluble fertilizer that supplies both calcium and nitrogen in a very soluble form of calcium to the roots or may be used as a foliar spray.

Also, there are a number of brands of liquid calcium or liquid lime. These are also quickly absorbed by the roots.

➤ Ground eggshells, fish bones, and seashells added to garden soil will supply calcium as they break down over the gardening season.

➤ Keep the garden soil evenly moist by using a thick layer of organic mulch and watering regularly.

General Discussion

Most planting mediums have adequate amounts of calcium. However, calcium should be added to the planting mix if the pH is too low. See the pH section for more information. If the water contains more than 150 parts per million (ppm) dissolved solids, it is probably providing the plants with enough calcium. If the water contains less than 150 ppm of dissolved solids, Ca-Mg (calcium-magnesium) has to be added to the water. To find out how hard or soft your water is, you will need to have a tds/ppm meter or refer to the local water district quality report.

Copper (Cu)

Copper deficient symptoms in tomato

How common is this deficiency?

Copper deficiency is occasionally a problem in home gardens. It is most likely to occur in fruit trees and sandy or low-organic content soils.

Symptoms

Copper deficiency first appears in young leaves that exhibit necrosis and coppery, bluish, or gray with metallic sheen coloring at the tips and margins. The young leaves turn yellow between the veins. The leaf petioles curl downward, and the leaf edges roll upward. Other symptoms include limp leaves that turn under at the edges and eventually die, and wilting of the whole plant. New growth has difficulty opening, flowers do not mature or open in males, and in females the stigmas don't grow properly.

Other symptoms include a tendency for the plant to wilt easily, stunted growth, dieback of stems and twigs, and yellowing of leaves.

Mobility

Copper has low mobility.

Role in Plant Nutrition

Copper is essential for carbohydrate and nitrogen metabolism. It is also required for the plant to form strong cell walls that guard against wilting. It is also used for flowcring and seed production.

CORRECTION

Typical symptoms of copper deficiency are a purplish or grey metallic sheen and leaves curling upward.

➤ Foliar feeding with copper fungicides such as copper sulfate and chelated copper can correct a deficiency. Micronutrient formula solutions containing copper work well. Natural sources are greensand, kelp concentrates, and compost.

Unusual grey sheen and twisted leaves are early signs of copper deficiency of this corn plant.

➤ Soaking dimes and quarters in water for a time (this is an inexact science), and then using the water to irrigate the plants, can also supply copper as these coins are 92 percent copper and 8 percent zinc. An acid solution dissolves the Cu. Interestingly, pennies are a poor source as they are mostly zinc and not Cu.

➤ Copper toxicity is rare but fatal. As the plant approaches death, its leaves yellow from its inability to use iron. The roots are abnormally sized, then start to decay.

Iron (Fe)

Severe deficiency in a maple leaf. The veins remain green but the tissue turns bright yellow.

How common is this deficiency?

Iron is usually abundantly available in soils. Iron deficiency occurs in areas where the soil pH is too high (pH 7.0 or higher) and where high levels of calcium carbonate exist (calcareous soils). High soil pH binds up the iron in a form unavailable for plants to utilize. Iron deficiency can also occur in areas where the topsoil has been removed or where there is scant organic matter and waterlogged soils. Foundation plantings growing near buildings are often iron deficient because the lime leaching out of the concrete foundations raises the soil pH.

It also appears at times in container gardening and hydroponics.

New tomato growth: The yellowing begins in the center of the leaf and works its way to the edges.

Symptoms

Iron deficiency starts in the new leaves, which lack chlorophyll but have no necrotic spots. This causes them to turn bright yellow except for the veins, which remain green. New leaves start to experience chlorotic molting, first near the base of the leaflets, so the middle of the leaf appears to have a brown mark. The veins remain dark green. Note that an iron deficiency looks similar to a magnesium deficiency except for its location. Iron deficiency affects the

Iron chlorosis is more likely to occur in acid-loving plants such as citrus. This branch is new growth near the treetop.

new growth but not the lower leaves while magnesium deficiency affects the middle and lower leaves first.

Mobility
Iron moves slowly in the plant.

Plants Affected
Acid-loving plants like azaleas, blueberries, citrus, dogwood, holly, magnolia, oak, rhododendron, sweet gum, and white pine are most susceptible, but many other plants can display symptoms in soils with high pH.

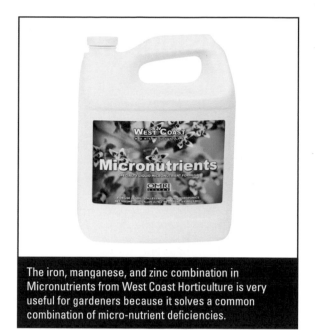

The iron, manganese, and zinc combination in Micronutrients from West Coast Horticulture is very useful for gardeners because it solves a common combination of micro-nutrient deficiencies.

Role in Plant Nutrition
Iron is essential for enzymes to function and acts as a catalyst for the synthesis of chlorophyll. Young, actively growing plants need iron to function normally.

General Discussion
Iron deficiency is sometimes found in combination with manganese and zinc deficiencies.

CORRECTION

➤ For a quick, short-term solution, foliar-feed iron-deficient plants with a chelated fertilizer containing iron, manganese, and zinc.

➤ Use iron oxides (Fe_2O_3, FeO) and iron sulfate ($FeSO_4$) for fast absorption. Use them foliarly and to the planting medium. Rusty water also works.

➤ Lower the soil pH by amending with well-processed compost.

➤ Mulch acid-loving plants with pine needles or other evergreen needles.

➤ Spray plants with seaweed extract.

Magnesium (Mg)

Cucumber leaf with slight deficiency. The veins remain green but the tissue turns yellow. It appears first in middle leaves.

How common is this deficiency?

Magnesium deficiency is not common in outdoor gardens, but frequently occurs in containers and hydroponic gardens.

Symptoms

Magnesium deficiency starts in the middle leaves with interveinal chlorosis; the veins remain green while the rest of the leaf turns yellow. The leaves eventually curl up and then die. The edges of affected leaves feel dry and crispy. As the deficiency continues, it moves from the middle leaves to upper leaves. Eventually, the growing shoots change from a pale green to white. The deficiency is quite apparent in the upper leaves. At the same time, the stems and petioles turn purple.

The deficiency appears first in the middle leaves as shown in this Phoenix palm. Then it works its way up.

Mobility

Magnesium is partially mobile.

Role in Plant Nutrition

Magnesium is a constituent of chlorophyll. It is also a catalyst in the metabolism of carbohydrates and breakdown of enzymes.

CORRECTION

➤ Water-soluble nutrients containing magnesium fix the deficiency. Such nutrients are magnesium sulfate ($MgSO_4$, also known as Epsom salts) and Ca-Mg for fast absorption and dolomite lime/garden lime and worm castings for moderate absorption. In planting mixes, magnesium deficiencies are easily fixed using 1 teaspoon Epsom salt per quart of water. After the first treatment, treat with one-quarter dose with each watering. Calcium-magnesium can also be used. For fastest action, Epsom salts can be used foliarly at the rate of 1 teaspoon per gallon. Calcium-magnesium can also be used foliarly as directed. Dolomitic limestone contains large amounts of magnesium. It can be used to raise the pH of soils and planting mixes and supply magnesium at the same time.

General Discussion

Magnesium deficiency is one of the easiest nutrient deficiencies to diagnose and cure. It occurs more frequently when using distilled or reverse-osmosis water and tap water that has a low parts-per-million count.

Magnesium deficiency stunts growth and lowers yields. West Coast Horticulture Magnesium is formulated to supply magnesium, in a fully soluble form.

Manganese (Mn)

The leaf sides curl upward. New growth turns yellow between veins as old growth loses tissue and becomes netted.

How common is this deficiency?

Magnesium deficiency is not common in outdoor gardens, but frequently occurs in containers and hydroponic gardens. Manganese is usually present in the medium, even when the plants are showing a deficiency. It becomes unavailable in alkaline soils as they dry and in moist highly organic soils. Soils sometimes undergo a cycle of availability and unavailability.

New growth cucumber leaf has internodal chlorosis, looking somewhat like iron deficiency but the smaller veins remain green.

Symptoms

Manganese-deficiency symptoms begin as a light chlorosis of the tissue between the veins of new young leaves. This area grows, getting lighter until only the veins remain green. Necrotic spots develop in the yellowed area. The older leaves look netted—only the veins remain green. As the deficiency continues, leaves develop a light gray or purplish metallic sheen. Maturity is delayed. Too much manganese results in an iron (Fe) deficiency.

Mobility

Manganese is not mobile.

Role in Plant Nutrition

Manganese plays a large role in photosynthesis. It is a catalyst in the synthesis of chlorophyll. It helps plants assimilate carbon dioxide for sugar production. It also helps assimilate nitrogen and makes it available for protein production.

CORRECTION

➻ Foliar feeding with water-soluble fertilizers high in manganese, such as Fe-Zn-Mn fertilizers or manganese chelates, provides quick relief from a deficiency. After the initial foliar treatment, add the fertilizers to the water.

➻ Compost and greensand can provide manganese, but it is absorbed more slowly than the water-soluble formulations.

Molybdenum (Mo)

Deficient tomato. Leaves in the middle of the plant have a pale yellow scorched look and irregular growth.

How common is this deficiency?

Molybdenum deficiency is very rare because plants require very little of it, but it occurs when the soil pH is below 5.5, and it becomes unavailable. These conditions occur in acidic, sandy soils in humid areas. The uptake of molybdenum increases as the soil pH increases.

Symptoms

The middle leaves turn yellow. As the deficiency progresses toward the shoots, the new leaves become distorted or twisted. A molybdenum deficiency causes leaves to have a pale, fringed, and scorched look, along with retarded or strange-looking leaf growth. Older chlorotic leaves experience rolled margins, stunted growth, and red tips that move inward toward the middle of the leaves.

Leaves of sweet potato are dull green, look scorched and have reddened veins.

Sometimes Mo deficiency is misdiagnosed as a N (nitrogen) deficiency. However, N affects the bottom leaves first. Mo affects leaves in the middle of the plant first and then moves up to the newer growth.

In cauliflower and some other crucifers, the deficiency causes young leaves to fail to develop and unfurl properly, giving them a whiplike appearance appropriately called whiptail.

Mobility

Molybdenum is mobile.

Role in Plant Nutrition

Molybdenum is contained in enzymes that help plants convert nitrate nutrients into ammonia. Without it plants cannot utilize nitrates. It is required for protein production within the plant.

CORRECTION

➤ Check the soil or water pH. If it is low, raise it, and Mo will dissolve and become available. Foliar feeding with a water-soluble micronutrient fertilizer is effective in correcting this deficiency. Micronutrient formulas can be added to the soil.

Nitrogen (N)

How common is this deficiency?

Nitrogen deficiency is a common occurrence in gardens.

Symptoms

First, lower leaves appear pale green. The leaves then yellow and die as the nitrogen travels to support new growth. Eventually, the deficiency travels up the plant until only the new growth is green, leaving the lowest leaves to yellow and wither. Lower leaves die from the leaf tips inward. Other symptoms include smaller leaves, slow growth, and a sparse profile. The stems and petioles turn a red or purple tinge in some plants. Too much nitrogen causes a lush dark green growth that is more susceptible to insects and disease. The stalks become brittle and break from lack of flexibility.

Tomato leaf from bottom of the plant turned pale yellow. The stems turned purple.

The older leaves of this broccoli plant turn yellow as the Nitrogen travels to new growth.

Mobility
Nitrogen is very mobile. It travels to new growth.

Role in Plant Nutrition
Nitrogen is needed for the production of chlorophyll and amino acids, and it is essential for photosynthesis. It is a vital element of tissue; without it growth quickly stops.

It is a component of tissue so without it growth quickly stops.

CORRECTION

- Any water-soluble nitrogen (especially nitrates, NO_3) that is quickly available to the roots will correct this deficiency. Insoluble nitrogen (such as urea) needs to be broken down by microbes in the soil before the roots can absorb it. After fertilization, nitrogen-deficient plants absorb N as soon as it is available and start to change from pale to a healthy-looking kelly green. Deficient plants usually recover in about a week, but the most-affected leaves do not recover.

- Nitrogen is the first number of the three-number set found on all fertilizer packages, which list N-P-K always in that order. Any water-soluble fertilizer much higher in N than P and K can be used to solve N deficiencies very quickly. Calcium nitrate ($CaNO_3$) is water soluble and fast acting. It can be used as a foliar fertilizer and in a water/nutrient solution. Urine, fish emulsion, (5-1-1), and high-nitrogen bat or seabird guano also act quickly. In soil, high-nitrogen fertilizers such as alfalfa and cottonseed meals, manure, feather meal, and fish meal all supply nitrogen fairly quickly but release it over the growing season.

General Discussion
Without high amounts of nitrogen, especially during the first stages of growth, later yield is greatly reduced. N issues happen throughout the entire growth cycle, but it is especially critical during the first part of growth.

A small amount of N is always necessary in order for the plant to manufacture amino acids, which use N as an ingredient. This supports flower and fruit growth and utilization of P and K.

Too much N causes dark green, brittle leaves that are attractive to insects. Growth speeds up, but the branches are thin and weak.

Legumes have specialized structures on their roots called nitrogen-fixing nodules. These nodules are formed by the soilborne bacterium *Rhizobium*, which invades the roots and forms the nodule. The bacteria that reside inside the nodule change the gas N_2, which is a component of air that plants can't use, to NH_3—the nitrogen form plants use. A leguminous cover crop grown every other season leaves N in the soil and provides organic matter when plowed under.

pH

Most garden plants do well in a soil range of 5.8 to 6.5. On either side of this range, nutrients become less available even when they are present in adequate amounts. pH values also affect the interaction between plants and soil bacteria or fungal diseases, either enhancing or inhibiting the ability of these organisms to infect plants.

Some plants prefer a lower pH range and acidic mediums. Plants such as azalea, camellia, citrus, lilac, and rhododendron prefer a pH of 4.5 to 6.0.

The only accurate way to adjust the pH is by using a pH meter or pH test papers. Guesswork won't do. When the pH is outside the 5.8 to 6.5 range, nutrients do not dissolve well and are not as available to plants. These are exceptions. Some plants prefer more acidic soils. As a result, though nutrients are present, roots do not have access to them so deficiencies develop. Plants growing outside the proper pH range grow very slowly and have small, dark green leaves. If the plants are growing in soil or planting mix, check pH using a pH meter or test strips before you plant. To do this, used collected run-off water. Adjust the pH using soil sulfur if it is too alkaline or lime if it is too acidic. Check with a knowledgeable local nursery staffer or agricultural extension agent familiar with local soils. He or she can give you advice on proportions since soils vary in their reaction to adjustments.

Hanna's pH Checker measures the water pH, which is important to optimize fertilizer utilization and promote fast plant growth.

Most indoor planting media are not soils at all: they are made using bark, peat moss, or coir as main ingredients, and other materials are added to adjust porosity and water retention. These mixes can be considered disease and pest free. Most commercial potting soils and topsoil are already pH balanced. If the plants are already in the ground and the soil is out of the preferred range, adjust the irrigation water using ph Up, liquid lime, wood ashes, or calcium nitrate to raise alkalinity, or liquid sulfur or pH down to increase acidity. Recheck the soil pH from time to time to judge when to stop these treatments. Hydroponic solutions should be kept in the 5.6 to 6.3 range using ph Up or pH down. Plants vary a bit in their ideal pH levels. Water should be pH adjusted after soluble fertilizers are added to it because their ingredients affect water pH.

How common this problem?

Problems with pH values are a common garden problem and can vary from location to location and with different types of planting media and different types of plants.

Symptoms

The symptoms exhibited by plants suffering from pH imbalance are those of nutrient deficiencies because the pH controls how readily these nutrients are available. For instance, in acidic soils the availability of calcium (Ca), magnesium (Mg), and potassium (K) is reduced while the availability of nutrients potentially toxic to the plant, such as aluminum (Al), iron (Fe), and zinc (Zn), are increased. On the other hand, alkaline soils cause boron (B), iron, manganese (Mn), and zinc to be less available to the plant.

CORRECTION

➤ To determine the pH of soil, use a pH meter or a soil testing kit available at garden supply centers, or contact the local agricultural extension agency to arrange for testing. Guessing won't work.

➤ Correcting the pH of acidic soils is achieved by adding lime in recommended amounts. Lowering the pH of alkaline soils is more difficult. Adding elemental sulfur allows soil bacteria to convert this sulfur into sulfuric acid, thereby changing the pH of the soil. Planting landscape and gardening plants suited to the existing soil pH is easier than making drastic changes to the soil conditions. Contact a knowledgeable garden center or agricultural extension agent familiar with local soils for advice and follow any recommendations given by the testing facility.

Phosphorus (P)

Midrange cabbage leaves turned a bluish-purple color and then develop brown spots. Older leaves turn yellow. The plant is stunted.

How common is this deficiency?

Phosphorus deficiency is uncommon.

Symptoms

Phosphorus deficiency is difficult to identify visually and can be mistaken for other deficiencies such as nitrogen and zinc. The most common indication is slow growth, stunting, or dwarfing—plants that grow slowly and are mistakenly thought to be much younger than they are. Leaves develop a bluish or purplish cast. Edges of older leaves have necrotic tan or brown spots, netted veining, and curl downward as the deficiency works its way inward. The lower leaves turn yellow and die. The stems and petioles turn purple or red. Flower and fruit production is inhibited in plants deficient in phosphorus. Plants use high amounts of P during flowering. If they don't get adequate or even abundant supplies, lower yields will result.

Tomato plant leaves turn purple and the sides of the leaves twist down.

Mobility

Phosphorus is immobile in the soil. Plant roots must grow to reach it, but it is mobile in the plant. Soil temperature plays a large role in the availability of phosphorus and in uptake by the plant.

Role in Plant Nutrition

Phosphorus aids in root and stem growth, influences the vigor of the plant, and helps seedlings germinate. Phosphorus is extremely important in flowering.

CORRECTION

➤ Phosphorus is the second of the three-number ratio listed on fertilizer packages. Water-soluble fertilizers containing high phosphorus fix the deficiency. Bloom fertilizers are high-P formulas. High-P guano also provides readily available P. Rock phosphate and greensand are also high in P and gradually release it. The affected leaves do not show recovery, but no additional growth is affected, and new growth appears healthy.

➤ Rock phosphate is treated with acid to create powerful, readily available nutrients. Triple superphosphate, diammonium phosphate, and monoammonium phosphate are all readily available soluble fertilizers but are not considered organic.

General Discussion

Deficiency during flowering results in lower yields, but overfertilizing can burn plants.

Cold weather (below 50°F/10°C) can make phosphorus absorption very troublesome. For this reason, soluble P, such as found in water-soluble bloom formulas, can aid flower yield in cool weather.

Potassium (K)

First the tomato leaf develops browned necrotic areas along the perimeter. Then the necrosis spreads over the leaf.

How common is this deficiency?

Potassium deficiency occurs occasionally in planting mixes and outdoors. Improper fertilization can result in deficiencies even in rich, well-fertilized soil.

Symptoms

Plants suffering from minor deficiencies look vigorous, even taller than the rest of the population, but the tips and edges of their bottom leaves die or turn tan or brown and develop necrotic spots. As the deficiency gets more severe, the leaves develop chlorotic spots. Mottled patches of red and yellow appear between the veins, which remain green, accompanied by red stems and petioles and slower growth. The leaves grow smaller than usual and larger older leaves develop dead patches, or necrosis, on their margins. These leaves eventually turn brown and die.

Rarely, excess potassium causes mature leaves to show a light to dark yellow or white color between the veins.

Ginger plant with deficiency. As it continues the entire plant is affected with failure to thrive and remains stunted.

Mobility

Potassium is mobile.

Role in Plant Nutrition

Potassium is found in the whole plant. It is necessary for water transport, as well as for all stages of growth; K aids in creating sturdy and thick stems, disease resistance, water respiration, and photosynthesis. It's especially important in the development of thick flowers and fruit.

Although symptoms of minor potassium deficiency affect the cosmetic look of the plant, it does not seem to affect plant growth or yields.

Nutri-K's potassium chemistry keeps it soluble so it remains available to the plants. It prevents deficiencies and eliminates them quickly.

CORRECTION

➤ Water-soluble fertilizers containing high potassium fix the deficiency. Bloom fertilizer usually contains high potassium levels. Wood ashes deliver K quickly. Kelp powder, liquefied kelp, bloom fertilizers, and wood ashes are commonly used and work quickly to correct K deficiencies. Potassium bicarbonate ($KHCO_3$), potassium sulfate (K_2SO_4), and potassium dihydrogen phosphate (KH_2PO_4) also act quickly. Potassium silicate (K_2SiO_3) can be used to supply Si (silicon) and has 3 percent K in it.

➤ Granite dust and greensand take more time to get to the plant and are not usually used to correct deficiencies, but for prevention. Damaged leaves never recover, but the plant shows recovery in four or five days with applications of fast-acting products.

General Discussion

Cold weather slows K absorption, as does too much Ca or NH_3 (ammonia). High levels of Na displace K.

Silicon (Si)

How common is this deficiency?

Silicon deficiency is very rare. Silicon is the most abundant mineral in the earth's crust. Its role as an "essential element" for plant growth is undetermined. However, plants grown with it have more resistance to disease, drought, and pests. It is deposited in the cell wall and in spaces between the cells in stems and leaves.

Symptoms

Silicon helps plants develop strong, stress-resistant leaves, roots, and stems. As a result, plants that lack silicon appear unhealthy, prone to repeated invasions by bacterial and fungicidal diseases and insect infestations. Silicon increases plants' photosynthetic activity and yields.

Mobility

Silicon is not mobile.

Role in Plant Nutrition

Silicon helps plants overcome stresses and provides resistance to pests, diseases, and stress by strengthening stem and branch structure. It also aids in growth, development, and yields.

CORRECTION

Diatomaceous earth

Greensand

Potassium silicate (sold as a liquid suitable for foliar application)

Silica planting medium

Sulfur (S)

Tomato with deficiency. The entire plant turns yellow or orange/yellow. The leaves have necrotic areas near the petiole.

How common is this deficiency?

Sulfur deficiency is rare.

Symptoms

Leaves exhibit an overall chlorosis while maintaining some green color. Veins and petioles turn a distinct, reddish color. Sulfur deficiency may resemble nitrogen deficiency, but with a lack of sulfur the yellowing is more widespread over the plant, including the young leaves. The reddening of the veins, petioles, and undersides of the leaves is less vivid than that exhibited with nitrogen deficiency. As the deficiency advances, leaves may become brittle, more erect, or twisted. Necrotic lesions develop along the petioles.

Overall growth is stunted. Sometimes S deficiency indicates as orange and red tints rather than yellowing. In severe cases the veins of the growing shoots turn yellow with dead areas at the base of leaves. The stems become hard and thin and may become woody. They increase in length but not in diameter.

Corn plants without adequate amounts of sulfur remain stunted and yellow.

Too much S stunts the plant and leaf size, and the leaves look brown and dead at the tips. An excess of S looks like salt damage: restricted growth and dark color damage. This is also rare.

Mobility

Sulfur moves slowly in the plant. The warmer the temperature, the harder it is for the plant to absorb sulfur.

Role in Plant Nutrition

Sulfur is essential to vegetative growth and is also important in root growth, chlorophyll supply, and plant proteins.

CORRECTION

➤ Both organic soils and inorganic fertilizers contain high levels of available S so plants are not likely to suffer from a lack of the element. However, a deficiency is easily solved using Epsom salts ($MgSO_4$). Water the plant with Epsom salts until the condition improves. Mix 1 teaspoon of Epsom salts per gallon and apply both foliarly and to the irrigation water.

➤ Adding water-soluble fertilizers containing S fixes the deficiency. Other nutrients are often mixed with sulphates. Mix at recommended strength to avoid nutrient burn. Other sources are elemental garden sulfur, potassium sulfate (K_2SO_4), and gypsum. Do not use gypsum on acidic soil (pH less than 5.5); it affects the absorption of soil aluminum, which is poisonous to plant roots.

Zinc (Zn)

Deficient tomato leaf is twisted, has turned yellow and then dies and turns brown.

How common is this deficiency?

Zinc deficiency occurs occasionally. It is often associated with iron (Fe) and manganese (Mn). They are less available to plants when the pH is high.

Symptoms

New growth has short internodes. Leaves

The leaf ends of deficient onions twist, yellow and quickly turn brown. The symptoms progress quickly.

are radically twisted with decreased size that has a rosette appearance in some species. Older leaves exhibit pitting in the upper interveinal surfaces. As the deficiency progresses, the interveinal yellowing turns brown while the veins remain green. Leaves of some species, including corn, grains, and grasses, exhibit yellow streaking between the veins of older leaves, or spotting. Interveinal yellowing is often accompanied by overall paleness. Flowers may contort and twist. When the deficiency first appears, the spotting can resemble an Fe or Mn deficiency, but it affects the new growth.

Zn excess is very rare, but it produces wilting and even death in extreme cases.

Mobility
Zinc is not mobile, and symptoms appear primarily in new growth.

Role in Plant Nutrition
Zinc plays a part in determining plant size and maturation. It is essential for the production of leaves, stalks, stems, and branches. Zinc is also an ingredient of many enzymes, including the growth hormone auxin. It is also required for the formation and activity of chlorophyll. Plants with higher levels of zinc can better tolerate drought conditions.

CORRECTION

➤ A Fe-Zn-Mn micronutrient mix solves the deficiency.

➤ Zinc salts, zinc sulfate ($ZnSO_4$), chelated zinc, or zinc oxide (ZnO) corrects the problem.

4. ENVIRONMENTAL STRESSES

Temperature, humidity, air quality, and the amount and type of light all affect your garden's health and yield. Whether plants are growing indoors or outdoors, the gardener must do as much as possible to optimize the environment that they grow in. The problems discussed in this section are mainly caused by grower error. You get details on how to fix each mistake and how to avoid its recurrence.

When you match the environment and plant needs, the result is a healthy, vigorous good-looking plant with profuse flowers or high yields. Luckily, there are plants that match virtually any indoor or outdoor situation.

If your plant or garden is suffering, at least one of the environmental conditions is out sync with its needs. Indoors, the situation may be corrected by moving the plant to another location. Light and temperature conditions change seasonally indoors so the plant or garden might be relocated or at least repositioned. Outdoors, the light pattern also changes seasonally. In some gardens some plants do better placed in containers so they can be moved to catch the right light.

Make sure you know a plant's requirements and preferred conditions when considering acquiring a plant or attempting to grow a crop. Choose only the species that will do well under the conditions you provide. This strategy will save you a considerable amount of time and effort and help you grow a great garden.

Container Conditions

Problem: Root-bound plants

Plants become root-bound when their roots have reached the perimeter of their container and start circling. There are two problems with this. First, being close to the edge of the container, the roots are subject to dramatic changes in temperature: burning from the sun, and getting chilled at night. The other problem is that when planted into a larger container, the roots have a hard time breaking out of the circle.

SOLUTION

➤ Root binding is an indication that the plants are in too small of a pot, so they should be transplanted. When transplanting, try to loosen the roots some, so that the root tips, where they grow from, can reach the new planting mix. To prevent this from happening, use a root pruning pot. These growing containers, such as the Smart Pot Pro from High Caliper, discourage the roots from growing near the edge and instead they fill the media throughout the container.

Problem: Dry or anaerobic conditions

SOLUTION

Smart Pots offer soft container solutions

➤ The planting mix that is used should be adjusted depending on the size of the container and the kind of plant growing. When dense planting mix with small particles is used in a large container, it compacts tightly, holding water and not allowing air to the roots. Planting mix with larger particles should be used.

➤ When large-particle soil is used in small containers, it drains very rapidly, creating dry conditions. Either the containers need frequent watering, or a denser planting mix.

Garden Room Conditions

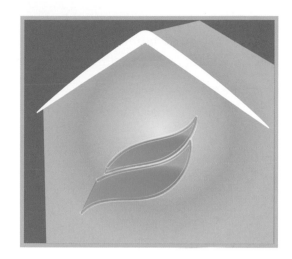

Problem: The greenhouse is too hot

SOLUTION

The following are a few ways to cool down a hot growing space:

► It can be ventilated to remove the heat.

► An air-conditioner can be used to remove the heat from the air.

► Use an air cooler that evaporates water to cool the room. It does not add too much moisture and can be used as long as the moisture level stays within limits. Air coolers are very efficient.

► If your garden is lit entirely using lights and is too hot during the day, and there is a big drop in temperature at night, many of the heat problems may be solved by running the garden lights at night, rather than during the day.

► If the plants are grown hydroponically, the water can be cooled. This keeps the entire plant cool even if the air is hot. If the room temperature is in the 80s°F (27°C to 30°C), keep the water temperature in the 60s°F (15°C to 20°C). Aquarium water chillers can be installed in the system to cool the water as it passes through the line.

► Remove all heat-producing equipment from the growing space if possible, especially light ballasts. If you use lights to illuminate your garden, you can prevent their heat from getting into the garden to begin with. Using air- or water-cooled lights keeps the heat generated by the lamps from entering the garden. Tubes carry it from the lamps to the outside without affecting room air temperature.

Problem: The growing space is very humid

SOLUTION

► The garden space is humid because the plants emit moisture and it is accumulating in the garden space. The moisture must be removed. The easiest way is to ventilate it out and replace it with drier air. If the garden is a closed system, you can use a dehumidifier. This appliance features cold tubing that condenses the moisture from the air. The problem with dehumidifiers is that they release heat and can make the space too hot. Air-vent tubing can be affixed to the exhaust so the hot air is directed out of the space.

➤ If the room is already running hot, an air conditioner can be used to remove the moisture as it cools the garden.

Herbicide Injury

Problem

Herbicide injury is common in areas where herbicide usage is necessary or frequent. Virtually all plants can be affected depending on the type of herbicide used. This injury occurs most often when spraying is done too close to nontarget plants or in breezy conditions that result in drift.

Damage can affect leaves, stems, roots, and whole plants depending on the type of herbicide used and the degree of coverage. Leaves may be cupped, distorted, or twisted and may turn white or yellow. Depending on the type of herbicide used, leaves may also turn very dark green and leaf margins die. Stems twist and become misshapen along with fruit and flowers. Prematurely dropped fruits and aborted buds may also occur as well as stunted root development. Once a plant is sprayed with an herbicide, there is little that can be done to save the plant.

SOLUTION

➤ Avoid spraying herbicide during breezy or windy conditions.

➤ Plant gardens or other plantings well away from any areas regularly sprayed with herbicides. Avoid spraying herbicide too close to nontarget plants.

- Spray on bright, sunny days to ensure that the herbicide dries as quickly as possible.
- Ask neighboring landowners to limit herbicide use near property boundaries.

Nutrient Burn

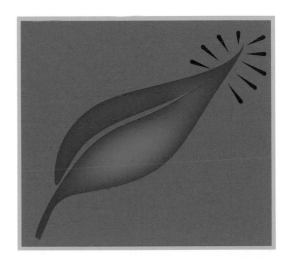

Problem

Very deep green leaves indicate an over-abundance of nitrogen. Burned tips may indicate too much potassium. Severe over-fertilization results in wilting.

SOLUTION

- Stop fertilizing. Flush the medium with pH-adjusted nonfertilized water. Sometimes plants appear to be suffering from a nutrient deficiency, but treatment fails to solve the problem—it could be an excess of nutrients. This creates chemical reactions that lock up nutrients, making them insoluble and preventing them from being absorbed by the roots. When this occurs, flush the plant with twice the amount of water as the size of the plant container.

- No two nutrient burns are alike; they have different symptoms and mimic different problems. If you have added nutrients and the problems persist, flush the plants and lower the nutrient concentration. Feed plants according to their size. Large plants use more nutrients than smaller ones.

- It is always best to dilute fertilizer solutions and gradually increase concentrations. If you add too much, you may have to flush.

Ozone and PAN

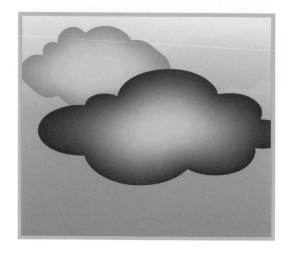

Problem

Ozone damage is a common problem in gardens in or near highly urbanized areas. Ozone is produced when sunlight reacts with automobile exhaust gases. On hot, calm days in mid- to late summer, ozone pollution can reach damaging levels. PAN (peroxyacyl nitrate) damage is found in smoggy areas. PAN is also a product of car exhaust.

Ozone and PAN damage is most often seen on the leaves of plants. Ozone damage appears as white-to-tan stippling on leaves. Plant growth is stunted, and flowering and fruit production is depressed. High levels of ozone result in early fall color and premature leaf drop.

A wide range of plants can be affected. Sensitivity varies greatly among species of plants. Particularly sensitive to ozone are asters, beans, begonias, blackberries, celery, fuchsia, lettuce, peppers, petunias, pines, spinach, sweet gum trees, tomatoes, and tulip poplars.

Least likely to be affected are cabbage, cucumbers, English Ivy, snapdragons, squash, and sugar maples.

Young, rapidly growing plant tissue is the most likely to be affected by PAN exposure, although a wide variety of woody and herbaceous plants are susceptible as well.

Overexposure to PAN causes a silvery glaze on the lower surface of leaves, giving them the appearance of having frost or sunscald damage, or an infestation of leafhoppers, mites, or thrips.

SOLUTION

➤ There is no cure for either ozone or PAN damage. Avoid growing sensitive plants in areas likely to experience ozone damage in summer.

Salt Injury

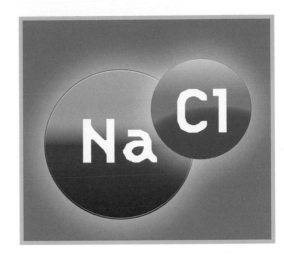

Problem

While some soils are naturally high in salts, salt injury most commonly occurs in areas where salt is used for de-icing during the winter months and in houseplants where watering may not be adequate to leach out accumulated salts. In the north, salt (sodium chloride [NaCl]) used for de-icing remains in the soil after application to roads, driveways, and sidewalks in winter. In areas close to the ocean, sea spray can harm plants. In the western states, soils and irrigation water are often naturally high in salt. Excess use of fertilizer causes a buildup of salt in soil when drainage is poor or when watering or rainfall is insufficient.

Almost all plants unaccustomed to salty conditions are affected. Salt stress affects the entire plant but may be most visible on the side of the plant facing a walkway, driveway, or road that has been de-iced. Leaves may have brown tips and dry out and wilt. Whole plants may wilt or appear stunted. A white, crusty material appears on the surface of potting soil when houseplants are underwatered and salt is not leached from the soil. The buds of plants subjected to salt exposure lose their winter hardiness and develop a sensitivity to the cold. Repeated exposure to excess salt can result in an overgrowth called a witch's broom effect.

SOLUTION

➡ Water heavily to remove excess soils from outdoor plants exposed to salt from de-icing, and from the potting soils of houseplants.

➡ Erect barriers of plastic between plants and areas likely to be de-iced.

➡ Use sawdust, sand, or nonsodium-based compounds for traction or de-icing near planted areas.

➡ Improve soil drainage to facilitate the leaching of salt by adding perlite or sand to potting mixes.

➡ In the west, where soils are high in salt and drainage is poor, add gypsum to the soil.

➡ For areas where salt exposure is unavoidable, choose salt-tolerant plant species and cultivars such as black currant, crack willow *(Salix fragilis)*, European larch *(Larix deciduas)*, Mugo pine *(Pinus mugo)*, pear, sycamore maple *(Acer pseudoplatanus)*, Virginia creeper *(Parthenocissus quinquefolia)*, and white fir *(Abies concolor)*.

In the United States each winter, more that 15 million tons of salt are applied to driveways, roads, and sidewalks.

Soil

Clay and rain-saturated soils are the usual causes of poor drainage. Sandy soils, on the other hand, may drain too well.

Problem: Clay Soils

Clay soils are notorious for drainage problems. They form a virtually nonporous layer so the water just puddles. This creates an anaerobic condition that damages the roots.

SOLUTION

- Before planting, dig up the clay. Modify soils that are composed of clay but with substantial amounts of other materials, using compost, fresh organic matter, sand, perlite, or used planting mix. Gypsum and sulfur are sometimes used to modify clay chemically. They both act to break up its tight molecular structure.

- Soils that are almost all clay are very difficult to garden. Alternatives are creating a raised bed or using a large container to hold soil aboveground, or excavating the clay and replacing it with different soil. If you are digging planting holes, try to reach a permeable layer. You might use an auger bit on a drill to create a drainage hole that reaches a permeable layer. If you know that an area will be saturated from rain during the growing season, build mounds or raised beds to keep the roots comfortably above the water level.

Problem: Dried-out Soils

SOLUTION

➡ Add a wetting agent to the water to prevent it from beading on the surface of the soil. Wetting agents allow water to be easily absorbed into the soil and are available at garden shops. Gardeners sometimes use soap such as Dr. Bronner's or detergents as wetting agents.

Rootbound plant

Problem: Sandy Soils

Sandy soils don't hold water. Instead water drains, quickly leaving the plants thirsty if it isn't supplied quite often, sometimes even several times a day.

SOLUTION

➡ To increase the soil's water-holding qualities, add compost and other decayed plant matter. This may require moving a lot of material and may not be feasible. Water-holding crystals also help.

➡ Irrigate over a period of time using small amounts of water such as a drip. This way the ground is kept evenly moist. When it is watered all at once, it drains and after a short time the soil becomes moisture deficient. One simple solution is to use a 2½-gallon (9.5 liter) or larger water container that has an adjustable spigot. Adjust it to a steady drip, so it drips the whole day. Those steady drops will keep the plant from getting thirsty. Refill the container when it is empty.

➡ Another solution is to dig a planting hole 18 inches (45 centimeters) deep. Put a plastic tray 3 to 6 inches (7 to 15 centimeters) deep or a heavy duty plastic bag at the bottom of the hole and fill it up with the soil. The tray or bag will act as an underground reservoir and keep water from draining too quickly.

Stretching

Problem

Plants with long, thin stems that stretch are weak and can barely support the top canopy. Stretching occurs most frequently when starting seeds. Older plants suffering from light deprivation stretch toward the light. Flowers and fruit grow smaller and develop slowly.

Lack of light causes stem elongation. Under low-light conditions, seedlings grow long, thin stems, attempting to reach the light. Heat also affects stem elongation. Plants grow longer stems as the temperature rises if it is not coupled with an increase in light intensity. Sun-loving plants grown in a low-light–high-temperature regimen often stretch.

Some plants are genetically programmed to grow long stems between sets of leaves. Usually, the stems are thick enough to support the growth. This may be inconvenient, but it is not stretching.

SOLUTION

➡ To stop stem stretch, take the plants out of the shade and put them into the sun. If they have already stretched, support them using wooden skewers. Once they have brighter light, the stems will fill out and will be able to support themselves.

OTHER SOLUTIONS

➡ Plants that experience strong breezes that create stem and leaf movement grow stronger, shorter, and stockier stems. When the wind bends the stem, it creates tiny tears in the tissue. The plant quickly repairs these tears by growing new tissue. Brushing and bending plant leaves and stems a couple of times a day also helps widen the stem.

➡ For plants that naturally grow long stems, pruning may be the way to deal with height. Removing the top of the main stem forces growth of the surrounding branches. These branches don't grow as long as the main stem. Bending the top branch so it hangs low also spurs the growth of the lower branches.

Temperature

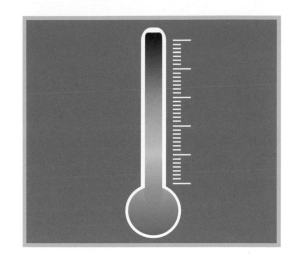

Problem: Low Temperature

When the temperature drops below 60°F, plant growth slows and yields suffer. This may not be noticeable if you aren't familiar with the garden's usual yield. Cool temperatures for a few nights won't make too much of a difference, but if they happen throughout the growth period, they can create a serious problem.

INDOOR & GREENHOUSE SOLUTIONS

➡ Most fruiting garden vegetables and summer flowers prefer moderate temperatures between 70°F and 75°F during the lit period, and a temperature drop of no more than 15°F at the lowest during the dark period. Plants enriched with CO_2 yield better at a slightly higher temperature, between 80°F and 85°F.

➡ If plants are growing in containers and are placed on cold ground or an unheated floor, heat will drain from the container. Use a layer of Styrofoam sheet to create a heat barrier.

➡ When the containers are kept warm, at 80°F, the roots will be heated and the stems and leaves will have more resistance to cold damage. Use plant-heating mats, which are placed under or in containers, to keep the planting medium warm. Aquarium heaters can also be used to keep the medium warm. Larger gardens can be heated using a hot-water heater set up to recirculate the water. The water should heat the planting mix no higher than 80°F. Use a CO_2 generator or an electric heater to maintain warm temperatures.

OUTDOOR SOLUTIONS

➡ Most fruiting plants can withstand temperatures as low as 50°F without problems. This is not an ideal temperature since it slows tissue growth at night and slows photosynthesis and growth when it occurs during the day. Temperatures below 40°F often result in tissue damage.

➡ Place row covers over plants while there is still warmth in the air to insulate the plants. Place filled water containers in the garden. They absorb heat during the day and release it at night, creating passive heating.

→ Gas patio heaters can keep plants warm on chilly nights. Keeping the temperature above 60°F increases plant growth.

Problem: High Temperature
SYMPTOM: ELONGATED STEMS AND DISTORTED FLOWERS

Plants can withstand high temperature as long as they have a large root system that can draw up enough water to keep the plant cool by transpiring water. If plants do not have a large root system, they will show elongated stems and distorted flowers.

INDOOR & GREENHOUSE SOLUTIONS

→ The air temperature in the aisles of a space isn't important; it's the temperature at the plant canopy level that affects the plants. Lower the temperature using ventilation, swamp coolers, or air conditioning. Eliminate causes of heat such as unprotected lights. Instead, use air- or water-cooled lamps.

→ If the floor is cool, place the containers directly on it to draw down heat. Cool the root temperature of the planting mix by recirculating cooled water in tubing under them. Use a water chiller to cool the water.

OUTDOOR SOLUTIONS

→ Shade plants during the hottest part of the day using agricultural shade cloth that provides 20 to 30 percent shading. This helps lower the temperature.

→ Cool the air using micro-sprayers that can lower the temperature 20°F. They emit a mist spray of tiny droplets that are 5 microns or smaller. The droplets evaporate, cooling the ambient air. These sprays are more effective when the air has low humidity. Both microsprayers and evaporating fans can be used.

→ Cover or paint black planting containers that heat up from sunlight. Use white or light-colored paint so light is reflected, keeping the containers and roots cool.

SYMPTOM: LEAF WILT

Plants with small root systems in relation to the canopy have a hard time drawing enough water to provide the canopy with water for transpiration, which is used by the plant to keep cool. This results in afternoon leaf wilt.

Make sure plants are provided with enough water to keep the roots moist. Give the container plant roots more room to grow by transplanting to a larger container. Cool plants using shade cloth or micro-sprayer misting.

SYMPTOM: TOP LEAVES APPEAR BURNED OR BLEACHED

The top leaves of plants look burned or bleached because they are growing in a space that is much too hot. The effect is often misdiagnosed as too bright a light. High-light-intensity plants can take the light—it's the heat that affects them.

Move the plants farther from the light. Install air-cooled lights with reflectors to reduce the heat near the light. Water-cooled lights are extremely effective at stopping light-generated heat from getting into the room.

Lamps can be placed closer when they are on a light mover. Setting it to move just a short distance changes the angle of the light in relation to the plants, but more important, it gives the top canopy a cooling-off period.

Plants that prefer more moderate light grow away from too bright a light. The leaves lighten in color as a result of producing less chlorophyll. The leaves may grow into weird positions to lessen the amount of light they receive.

Move the plant farther from the lightbulb or place it in a shadier area.

Usually in greenhouses, 1000-watt lamps should be kept about 3 feet (1 meter) from the plant tops. Air-cooled lights can be spaced 18 to 24 inches (45 to 60 centimeters) away. Water-cooled lights can be placed 12 inches (30 centimeters) or less from the tops.

Lights can be moved closer if they are placed on light movers. Plants that require more moderate light should be kept farther away from the light.

Water

Problem: Chlorine in Water

The irrigation water has large amounts of chlorine, which affects plants in several ways. First, in media systems it kills some of the microorganisms that form a community with the roots in the rhizosphere. This results in slower growth. Some leaf tip burn may be the result of excess chlorine. This is most likely to happen during hot sunny days when the plant uses a lot of water.

SOLUTION

Most gardeners water straight from the tap, which most likely has been treated with chlorine and possibly fluoride. In spite of these chemicals in the water, plants grow well. But they may grow better with nonchlorinated water. Most city water is chlorinated with chloramines, which cannot be removed by boiling or allowing the water to stand. If you wish to dechlorinate tap water, or chloramine-treated water, add a tablespoon of vitamin C, ascorbic acid, for every 75 gallons water. There are also specialized UV systems designed specifically for dechlorination. A number of chemical dechlorinators are available for use in fish ponds and aquariums. These products are safe for fish and plants; however, the author has not tried them and cannot say for certain which works best in the garden.

Problem: Drowning Roots/Overwatering

Plant roots require oxygen, not carbon dioxide. A lack of oxygen, called an anaerobic condition, causes the roots to lose their vigor, making them susceptible to attack by pathogens. Usually, roots obtain needed oxygen from the air spaces between soil particles. When these spaces become filled with water, the plants are deprived of oxygen. Symptoms are drooping but not wilted leaves, an unhealthy fatigued look, and very slow, if any, growth.

INDOORS & GREENHOUSES

There are three possible causes of drowning roots indoors:

1. You are watering too often and the planting mix does not have a chance to drain.

2. There are no drainage holes in the bottom of the container or the holes are clogged.

3. The planting mix's particles are too small for the container, compact together, don't allow enough air spaces between them, and thus hold lots of water.

In small containers, either fine-textured mixes hold moisture or the pots would dry out very quickly. Large-size particles are better for large containers because they don't become compact—then water can drain freely, leaving large air spaces.

SOLUTION

➤ Make sure the container has drainage holes. If not, cut some into the container using a knife, drill, or thermal tool. If the planting mix is composed of fine particles, water less frequently. Use a coarser, more aerated planting mix if you are transplanting and with future crops.

OUTDOORS

Outdoors, rain-saturated soils are the usual cause of poor drainage. Sandy soils, on the other hand, may drain too well.

SOLUTION

➤ See *Soil*, page 174.

Problem: Hard or Soft Water

Plants are suffering from erratic deficiencies and growth problems.

SOLUTION

➤ Check water for dissolved solids. It should be between 100 and 150 parts per million (ppm). Water that is too low (soft water) should be adjusted to 150 ppm using a calcium-

4H2O Zero Waste Filters remove 99+% of metals, calcium, and chlorine without wasting water, unlike reverse-osmosis filters, and balance the water's ionic charge, increasing fertilizer uptake.

magnesium solution. Water that is too high, with over 250 ppm of dissolved solids, should be filtered using ionic or reverse-osmosis filters.

➤ Hard water prevents balanced nutrient uptake. Filtered water can be mixed with unfiltered water to create water suitable for plants. Water in a Los Angeles suburb measured 450 ppm dissolved solids. After filtering through an ionic filter, it had a ppm of 13. The gardener mixed one part of unfiltered water with two parts of filtered water, resulting in a solids level in the mixed water of just above 150 ppm. The water has the right amount of dissolved solids for use in a nutrient-water solution.

Use pH Up or Down to adjust water pH

Problem: Sodium in Water

Avoid any water that contains high levels of sodium. Sodium is absorbed by the plant first, before any other element. It causes the plant's vascular system to break down. Water treated with water softeners should be avoided.

Problem: Sulfur in Water

Water with a sulfurous odor should be checked for its pH. Sulfur is acidic, and the water may have a very low pH. Adjust the pH of the water with ph Up. Soils that have been irrigated with the water may also be affected and have a low pH, indicating acidity. These soils should also be tested and adjusted if necessary using lime, which raises the pH.

When to Water

Roots of most ornamentals and vegetables should never dry out or the plant will wilt. There are several ways to determine when to irrigate. The first is touch.

➡ Use a finger to determine when the medium feels dry. Feel the mixture 2 inches below the surface of the soil level or at least 6 inches deep in containers.

➡ Use a water meter wand to determine moisture content. You can use it to measure levels at different depths.

➡ Containers can be measured by weight. Weigh the container right before you water and then again after it has been watered and drained. Water when the weight goes down. Remember, the container will weigh more as the plant grows.

The amount of water that a plant needs and how often it should be watered depends on the size of its roots and canopy, the water-holding capacity of the planting mix, light intensity and duration, temperature, and humidity.

➡ Larger plants require more water.

➡ Larger containers need water less frequently.

➡ Plants require more water when the temperature is warmer.

➡ Plants require more water when the humidity is lower.

➡ During the hot days of summer, soils often dry out quickly. Adding compost and water-holding crystals to the soil at planting time helps the soil hold water for longer periods.

➡ Covering the soil with mulch slows water evaporation dramatically. Use compost, wood chips, hay, dried leaves, newspapers, and rugs.

Hanna's Combo Tester checks pH and parts per million (PPM) of nutrients in water. It features options for advanced gardeners such as TDS measurement and adjustable EC/TDS conversion factors.

Weather

Problem: Cold Spells

The weather report says there will be a cold spell, with temperatures dipping below freezing. This is to be followed by warmer weather later in the month. How do I prepare my garden for the cold?

SOLUTION

➤ The idea here is to keep the plants alive until the weather changes. If the temperature can be kept at 45°F to 50°F or more depending on species, the plants will survive unscathed. Then when better weather returns, they will start growing again.

➤ If the plants can be moved inside and given moderate light on cycle, they can be preserved for a few days until the outside weather changes.

➤ A patio heater can be placed in the garden and may provide enough heat to keep the plant from frostbite. A temporary greenhouse constructed of wood frame and plastic will preserve the heat better and can be removed when better weather arrives.

➤ Individual plants can be wrapped using polyethylene. This protects the plants from wind and preserves some heat. However, the cold will eventually get to the plants unless there is some source of heat for them. Forced-air heaters can deliver heat to the plants. Make sure to set the gauge at about 70ºF so the plants do not overheat.

Problem: Cool Weather

The plants are not mature and the weather is getting cooler. How long can the plants stay outside?

SOLUTION

➤ Plant growth slows dramatically as the daylight temperature slips down into the low 60s°F, and virtually stops in the mid-50s°F. If it is unlikely that the temperature will rise to the high 60s°F or 70s°F, then it may be useless to keep the plants growing.

➤ At the same time, the evening temperature may be slipping dangerously low. Plants can withstand temperatures in the mid-40s°F, but when it slips into the 30s°F, tissue damage is likely.

- The solution depends on the amount of sunlight available. As fall turns closer to winter the intensity of sunlight diminishes tremendously and its light reaches earth at a lower, more oblique angle. Often plants that were in full light during summer and early fall are shaded most of the time by mid-fall. Clouds may obscure the sun most of the time. Plants do not receive enough light energy to support growth. They should be harvested.

- If plants are getting sunlight but are still experiencing cold weather, they could be protected using clear plastic hanging over a frame. The air inside the frame heats up, so growth is promoted.

- One way to keep the air warm at night in enclosed spaces is to use passive heaters made by filling dark-colored containers with water. The containers heat up during the day, then radiate heat at night.

- Patio propane heaters can also be used to keep the plants warm. Burning gas produces CO_2 and water vapor. The extra CO_2 promotes plant growth.

Problem: Rainy Weather

Rainy weather is forecast. How can the plants be protected?

SOLUTION

- Rainy weather promotes mold. Water gets into crevices where it is hard to dry out, creating a perfect environment for molds such as *Botrytis* to grow. If plants can be moved or a protective enclosure can be constructed, the plants will be protected from rain but not from moisture. Raising the temperature of the enclosed area into the high 70s°F, beyond the optimal range for mold growth, may protect plants and help dry out the moisture. A fan circulating the hot air helps with this task.

- If the rain is expected to be brief followed by warming dry weather, plants can be protected by treating them with an antifungal such as potassium bicarbonate or Serenade before the rain starts. If the forecast is for prolonged rain, the best solution may be to harvest the flowers and fruit.

Wilted Plants

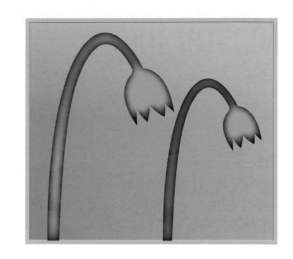

Problem: Wilting appears suddenly, but the soil is moist

One moment the plants are fine, and a few minutes later, wilting starts that can progress slowly or quickly. When the problem becomes severe the leaves remain green but dry in place.

SOLUTION

➤ Plants draw water up by keeping a higher salt concentration in their tissues than in the surrounding soil. If the salts (fertilizer nutrients) become more concentrated in the planting mix than in the plant, either the plant can no longer draw up water or it actually drains from the plant. Use pH-adjusted 72ºF (220C) water to flush the soil. Use water equal to about one and a half times the volume of the container, or about seven gallons of water to flush a five-gallon container. Reinstitute the fertilizer program after they have perked up. Leaves and stems that have dried and cannot be revived.

Problem: Plants grow slowly or are wilting

SOLUTION

➤ When water becomes scarce, wilting occurs, starting from the bottom and climbing to the top. If the amount of water is just barely adequate to maintain the plants, growth slows. This may be hard to notice. However, once plants receive more generous supplies of water, provided they have enough nutrients, they will experience a growth spurt.

Winter Injury

Problem

Winter injury is common in areas where temperatures often fall below freezing. The most common types of winter injury are frost damage, low-temperature damage, winter desiccation, winter sunscald, frost cracks, and snow and ice breakage.

Frost injury happens in late winter or early spring when active tissues are not hardened to the cold, such as early-emerging plants subjected to a late-spring cold snap. Low-temperature damage occurs when plants are grown north of their hardiness zones but can also occur in unusually cold conditions or when temperatures drop or change unexpectedly. Winter desiccation occurs in evergreens. Sunscald occurs during abrupt temperature variations when the sun heats and expands inner tree bark, then cold nighttime temperatures cool it down quickly. Frost cracks occur when the winter sun causes a differential expansion of the inner wood. Ice breakage occurs on weak limbs and limbs with foliate, like evergreens, when the snow load is too heavy.

Entire plants can be affected by winter injury, and damage can range from growing buds to whole plant death. Winter-damaged plants may live normally in spring, then die later when their stored food reserves are exhausted. Some winter damage may not be apparent for one to two years after the damage has occurred.

In broad-leaved plants, frost damage or low-temperature damage often appears as water-soaked spots or blotches. Evergreen needles turn yellow or brown with winter desiccation. Frost cracks form in the bark or wood of a tree during extreme cold, often accompanied by an audible loud crack, and may widen or narrow in response to temperature. These cracks often callus over in summer only to open again the next winter. Winter sunscald occurs most often in young trees or types of trees with thin bark. Ice breakage from snow and ice accumulation snaps branches off trees, especially conifers and evergreens with heavy winter foliage. A wide range of herbaceous and woody plants are susceptible.

SOLUTION

FROST DAMAGE

➤ Prune away damaged tissue. If the damage is not too severe, plants will recover.

LOW-TEMPERATURE DAMAGE

Winter injury on tomato

➥ Avoid using plants too far north of their hardiness zone. Containerized plants are especially susceptible as their root systems are not as protected as plants in the ground. Bring them inside, if possible, or else heavily mulch the pots, group them together, or sink them into the ground to provide protection. Avoid fertilizing heavily just before fall, especially fertilizers with high-nitrogen content, so plants do not produce tender growth that is susceptible to cold damage. Mulch around trees and plants to help protect their roots from the cold. Prune and discard any killed tissue to avoid invasion by disease organisms.

➥ Wrapping plant canopies in plastic and trunks in burlap helps.

WINTER DESICCATION

➥ Water plants well in late fall to early winter as sufficiently hydrated tissues can better withstand winter drying. Wrap susceptible plants in burlap to protect from winter winds. Mulch around plants to protect roots. Antidesiccant spray can help protect susceptible plants. Apply in late fall and again during winter.

➥ Spray with an anti-dessicant, such as Wilt-Pruf.

WINTER SUNSCALD

➥ Wrap the trunks of young or susceptible tress with a commercial tree wrap product to protect against winter winds.

FROST CRACKS

➥ Wrap tree trunks with plastic and burlap. Brace any cracks to facilitate healing and prevent the reopening of the cracks the next winter.

ICE BREAKAGE

➥ Prune evergreens in fall to reduce the amount of foliage available to collect snow and ice. Branches with narrow branch-to-trunk angles are more likely to break than those with broader angles. Do not place plants directly under roof drip lines. Wooden shelters can be built over valuable plants to protect them from ice and snow.

➥ Spray with a dilute solution of potassium bicarbonate ($KHCO_3$).

5. CONTROLS

In this section you will find descriptions of most of the control methods previously listed in this book. You will be reinformed on whether the control is used to battle nutrient deficiencies, pests, or diseases and will be given specific instruction on how to use each control. A wide range of solutions is presented, ranging from barriers, commercial products, homemade recipes, to bio-controls. Section 5 is divided into two alphabetized lists: Control Products and Beneficial Biologicals. Under Control Products you will find the active ingredients that make a product an effective solution. Some are derived from natural substances and others are extractions from plants and herbs.

The other category of controls, Beneficial Biologicals, consists of either living organisms or the products they produce.

All of the solutions suggested in this book are safe to use for herb or edible crops. Unless noted, they are totally pet and family friendly. These controls show you how to protect your garden safely.

A GENERAL NOTE ON USING PRODUCTS

Always read the label carefully before buying or using any control product.

Follow the manufacturer's directions carefully to use these products safely and avoid harming the plants. If you are applying any product that you have never used before, including a new homemade recipe, always test it on a few branches of several species of plants and wait a day or two before applying it to your entire garden.

Identify the ingredients in a product to be sure that it will treat the problem you're dealing with and will not introduce additional problems you don't want. Many modern garden products seek to be all-in-one cures for common garden problems—for example, a micronutrient fertilizer that also contains high levels of nitrogen. High levels of nitrogen might affect flowering.

Others apply more than one control for the same problem, such as a pyrethrum product that also contains rotenone or piperonyl butoxide (PBO). Check the label to avoid surprises. PBO is an adjuvant, which increases the potency and staying power of pyrethrum, but it is not organic.

Sprays are washed away by water, including rain, so plan to reapply spray products after rain or any watering that hits the affected areas of the plants.

Eco-Friendly Solutions

Alcohol Spray

A 50/50 dilution of 90 percent alcohol and water is an effective method of killing mealybugs and scale. Spray or dab directly on the bug weekly for five weeks, which disrupts the life cycle.

Alfalfa and Cottonseed Meal

These granulated products are made from pressed alfalfa hay and the solids left over from the extraction of oil from cottonseed. Both are high in nitrogen (N) held as protein. When either is used in planting mixes, it acts as a slow-release nitrogen fertilizer. Neither quickly remedies deficiencies but prevents them in the first place by releasing N gradually.

Ant Baits

These products contain a poison such as arsenic, boric acid, or sulfluramid and an attractive bait that entices the ants to carry it back to their nest, where it kills both workers and queens. Available in many brands, including Grant Ant Stakes, Terro Liquid Borax Ant Bait, Enforcer Ant Bait, Advance Dual-Choice Bait Station, and Drax Liquidator Ant Bait Station.

Avid

This product and others composed of abamectin are approved only for ornamentals, not edible plants. Do not be misled into using this product on plants intended for consumption. It is a nerve poison and affects the reproductive system.

Azadirachtin

This ingredient is made by processing a limonoid with neem oil in alcohol. Once ingested, insects are unable to feed, molt, or lay eggs, and soon die. It is relatively harmless to bees, butterflies, and other pollen-eating insects and to beneficial predators and parasites because it is effective only when ingested. It breaks down in five days in the presence of light and water.

Brands include AzaMax, Azatrol, Bioneem, and Neemix, among others. AzaSol is a water-soluble form of azadirachtin.

Baking Soda

Baking soda (sodium bicarbonate [NaH-CO$_3$]) can be used as a substitute for potassium bicarbonate—it works the same way, but as its close cousin, baking soda is often preferred by gardeners because it is so readily obtainable. However, it is not as effective as potassium bicarbonate and leaves sodium in the soil when it breaks down. It controls powdery mildew because it raises the pH of leaf surfaces to inhibit fungal growth. Rose growers have used sodium bicarbonate for powdery mildew for more than 70 years.

Recipe for Baking Soda Spray Solution

1 teaspoon baking soda

Few drops of castile soap or other wetting agent

1 pint water

Mix ingredients together in a spray bottle. Apply the spray weekly on new growth.

Black Pepper Spray

Black pepper spray is used both to prevent predation by caterpillars and to kill them. Black pepper contains irritants that disrupt caterpillar feeding and affect their skin.

Recipe for Black Pepper Spray

1 tablespoon ground black pepper

1 quart water

3 ounces potable alcohol such as 100-proof rum or vodka (optional)

Few drops wetting agent

The alcohol dissolves insect toxins not soluble in water. Mix and let stand for a day. Spray plants.

Boric Acid

Boric acid is a quick and effective cure for boron deficiency. Mix 1 teaspoon of boric acid, readily available in drug stores, with 1 gallon of water and spray on foliage.

Boric acid is also a potent ant, roach, and crawling insect killer. Insects crawl through the powder, which sticks to their skin. Then they lick it off. Once it is ingested, it acts as a poison and acid, destroying the insect's body internally. Boric acid is also used as an ingredient in various insect baits. It is often placed around the perimeter of indoor garden spaces to prevent invasion. It is harmless to pets and kids, but they should be discouraged from eating it.

Cal-Mag

Cal-mag supplements contain both calcium and magnesium and are effective for correcting deficiencies in either or both of these nutrients. It helps prevent blossom end rot in tomatoes and can increase the water and nutrient uptake in fruits and flowers. Rainwater and water gathered from seasonal runoff is usually mineral deficient, as is water that has gone through reverse-osmosis. Calcium and magnesium should be used as a supplement to bring water to 125 ppm dissolved solids before adding fertilizer. It is not needed if the ppm is already at that level.

Brands available include Botanicare

Cal-Mag Plus, Sensi-Cal, MagiCal, and Cal-Max.

Calcium Nitrate

Calcium nitrate ($CaNO_3$) is used to correct calcium deficiencies in the garden and is especially useful in that it supplies both calcium and nitrogen. It is highly soluble and provides these nutrients to the plants very quickly. It is a good solution to both Ca and N problems. It is available from a number of suppliers such as Viking, Vara-Liva, and Acros. It is not available at most gardening centers and garden shops but is readily available on the Internet.

Capsaicin

Capsaicin is the substance that pepper plants produce as protection against insects and gives hot peppers their heat. It repels insects from other plants and is also insecticidal by contact.

To make an insecticidal spray, use:

- *½ ounce dried or 4 ounces fresh habanero or other very hot peppers or 2 tablespoons Asian hot pepper oil*
- *2 tablespoons of vegetable oil*
- *¼ teaspoon lecithin granules*
- *¼ teaspoon wetting agent (like dish soap) Water to make one pint*

Always wear gloves when handling hot peppers or pepper oil and avoid touching eyes or mouth.

Grind the peppers, including the seeds, and oil in a blender. Add the lecithin granules, wetting agent, and water and mix thoroughly. The mixture can be used immediately, but when it sits for a time, it becomes stronger. Strain through cheesecloth or pantyhose and store in a glass jar. This makes a concentrate. To use, mix 1 ounce of pepper mix to 1 quart of water and spray. Test-spray a few leaves first and check after a day for damage before spraying the whole garden. Depending on the pepper potency, it may have to be used in a less concentrated mix.

Commercial products that contain capsaicin include Bonide Hot Pepper Wax (for ants, aphids, leaf miners, spider mites, thrips, and whiteflies); Repellex Mole, Vole, and Gopher Repellent (for moles and gophers); and Browseban and Liquid Fence (for deer).

Carbon Dioxide

Carbon dioxide (CO_2) occurs naturally in the air and is essential to all plants. In na-

The Green Pad provides CO_2 to indoor gardens, boosting growth.

ture, the level of CO_2 in the atmosphere is 380 parts per million (ppm). Plants quickly use up CO_2 in closed spaces so ventilation is essential for fast plant growth. Some plants grow even faster when the air is enriched with CO_2. This can be done using a CO_2 generator, which burns propane or natural gas, CO_2 tank and regulator, or using chemical means of producing CO_2. A product like the Green Pad generates carbon dioxide in indoor garden spaces. Spaces where people congregate generally have fairly high CO_2 levels.

Chamomile Tea

Chamomile blossoms contain sulfur, which suppresses damping off. Make a tea using ¼ cup of tightly packed chamomile blossoms by pouring a pint of boiling water over them and allowing the tea to steep until cool. Strain this liquid into a spray bottle. Soak seeds before planting, after the seeds have been planted, and when the seedlings emerge.

Chelated Minerals

Chelation is the process of combining nutrient minerals, such as boron, copper, iron, magnesium, manganese, zinc, and others with a compound such as citric acid or EDTA (ethylenediaminetetraacetic acid) to make them more available for uptake by plants. Hydroponic fertilizer formulas usually contain a blend of chelated minerals. Many standard fertilizers provide only NPK. Other products contain metal che-

lates for treating specific deficiencies. Usually, clay and loam soils contain adequate amounts of these micronutrients. However, sandy soils are often lacking. Some commercial brands are:

Copper: Librel Copper Chelate, YEOMAN 5% Cu

Iron: Bonide Liquid Iron (also contains zinc and manganese), Plant-Prod Iron Chelate, Librel Fe-DP

Manganese: Growth Product Manganese Chelate, Librel Manganese Chelate

Zinc: YEOMAN 7% Zn, Nulex Liquid Zinc

Cinnamon

Cinnamon contains cinnamaldehyde, which is a fungicide and a pesticide. It is used in the form of powder, oil, or tea. It has multiple uses in the garden. It is insecticidal and repellant to ants, aphids, fungus gnats, spider mites, thrips, whiteflies, and many other arthropods. It is also fungicidal and is used to kill both root and leaf invaders. Its effectiveness against powdery mildew ranges from 50 to 70 percent eradication and is most effective when used in combination with other ingredients. It won't eradicate the disease completely but will keep it from exploding. It is an ingredient in many oil blends, which potentiate each when used in combination.

With the recipes below, test on a branch first and wait 24 hours to make sure the mix is not too concentrated. If it is, dilute with water and retest. Damaged foliage looks brown and burned (necrotic). If there are just a few spots, dilute by 25 percent. If there are large necrotic areas, dilute by 50 percent or more.

Cinnamon Tea Recipe

1. Mix food grade cinnamon oil (available at herbal shops) at the rate of one part oil to 200 parts water (or 1 teaspoon to about 1 quart water). Add a wetting agent and ¼ teaspoon dry lecithin granules. Blend. The spray is ready to use.

2. Cinnamon tea: Boil 1½ pints of water. Turn off the heat and add 1 ounce of cinnamon powder. Allow the tea to cool to room temperature. Once cool, add 1 pint of 100-proof grain alcohol and water to make a quart. Strain the mixture through a filter to remove any bits of cinnamon. The spray is ready to use.

Some commercial products containing cinnamon are: Ed Rosenthal's Zero Tolerance, which contains cinnamon oil in both its fungicidal and pesticidal formulations; Dr. Earth Pro-Active Fruit and Vegetable Insect Spray; and FlowerPharm.

Citrus: see D-limonene

Clove Oil

Clove oil contains eugenol, which is a fungicide and a potent contact insecticide. It has residual activities as long as the scent lingers. Clove oil is effective against the fungi that cause damping off, *Fusarium*, gray mold, powdery mildew, *Pythium*, and *Septoria*, and many arthropods, including ants, aphids, beetles, caterpillars, mealybugs, spider mites, thrips, and other pests.

You can make your own oil or tea using the recipes shown under cinnamon. The two oils are more effective when used in combination.

Ed Rosenthal's Zero Tolerance contains clove oil in both the fungicidal and insecticidal formulations. Other products with clove oil include Dr. Earth Pro-Active Fruit and Vegetable Insect Spray; Phyta-Guard EC; Pest Out; Natural Planet Bug-A-Tak; and Bioganic Multi-Insect Killer Lawn & Garden Spray by Green Light.

Compost

Compost is the rich, black material created from the decomposition of organic matter such as leaves and other plant material, kitchen scraps, and manure. Many gardeners make their own using a compost pile or compost container. It is available at some municipal composting centers—which are often part of the waste-management system. Compost is also available at landscaping suppliers, garden shops, and garden centers. Adding compost to garden soil has several benefits:

➤ It increases the water-holding capacity of soil by adding organic matter.

➤ It provides beneficial microbes that act as barriers to infections and destroy plant pathogenic organisms.

- It increases the drainage ability of heavy soils such as clays.

- It contains many readily available nutrients held in the organic form.

Work compost into the top 2 or 3 inches of garden soil, or use as a side dressing or mulch around new plants or existing plantings. Compost can also be spread across an established lawn after aeration in a ¼- to ½-inch-thick layer.

Compost is an excellent ingredient in planting mixes. Use up to 25 percent compost in the mix.

Hot compost has been thoroughly heated during the decomposition process to temperatures that kill any pests or weed seeds. The heat is generated by the rapidly increasing population of beneficial microbes. A hot compost pile is hottest near the center, and these centralized temperatures should be between 130°F and 140°F for about a week. Temperatures lower than these allow pests and plant pathogenic organisms to survive the composting process. Temperatures higher than 150°F will kill the beneficial organisms creating the compost. Adding ingredients as the pile progresses helps keep the temperatures under control. Compost piles prepared in this manner may take a few days to reach these temperatures.

To prepare hot compost:

Gather materials such as grass clippings, kitchen scraps, newspapers, sawdust, garden waste, and manure from cows or horses. (Do NOT use dog or cat manure as their droppings may carry organisms harmful to humans.) Meat scraps and oils will attract nighttime foragers, including rats and raccoons, so if you use these ingredients, the pile should be constructed to prevent these animals from getting to it.

Chop up or shred large pieces. Tear up newspapers and chop up small branches and limbs to provide as much surface area as possible for microbes.

Combine the materials in a ratio of one or two parts dry, yellow, or brown plant-based (carbonaceous) materials to one part wet, green, or animal-based material. This translates to about equal parts leaves, sawdust, and newspaper with fresh grass clippings, food waste, or other high-nitrogen material such as manure. This is a general recommendation, and there is a lot of latitude in ingredient ratios. If no other high-nitrogen material is around, you can use ammonium nitrate or other high-nitrogen fertilizers to promote fermentation. Add several shovelfuls of soil or finished compost as you mix the ingredients to serve as a compost "starter." The soil will supply the needed microorganisms to start decomposition. Commercial starters are also available at gardening centers.

Turn the compost pile regularly, every few days if convenient, to supply oxygen to the organisms at work in the pile and to keep the temperatures at the center regulated. Compost thermometers are available to monitor the temperature.

When the original components of

the compost pile are no longer recognizable and the pile has cooled, the compost is finished.

Compost Tea

Compost tea is a dark brown liquid that is used foliarly as a plant stimulant to promote plant resistance to stress, and as a natural fungicide against the early stages of some diseases by introducing beneficial organisms, hormones, and enzymes. It can also be used as a soil drench.

Compost tea is an excellent means of promoting clone and seedling growth and protecting them from damping off and stem rot.

Compost tea is made by aerating a biological source such as finished compost, vermicompost, and Alaskan humus in water. Nutrients that promote microorganism growth such as humic acid and molasses are added. The aerated brew is ready to use in 24 hours.

Kits to produce compost tea are available at gardening centers. Some indoor garden shops sell the tea already brewed, ready to use.

A home brewing option is the Bountea Garden Tea Brew Kit

To make compost tea:

Assemble the needed materials: bucket, aquarium pump with hose and bubbler, nylon stocking, and enough compost to fill the bucket about ¼ full.

Put the compost into the stocking to make a "tea bag." Suspend the stocking in the bucket by tying the end to the bucket handle.

Add the nutrients that promote microorganism growth such as molasses and humic acid.

Fill the bucket with water and put the air hose and bubbler in the bottom of the bucket. Turn on the pump and aerate for a day. Switch off the pump and let the solution settle.

The resulting liquid should be dark brown and have a pleasant earthy smell. An ammonia or rotten odor indicates the tea is anaerobic and should not be used as is. Strain the liquid through cheesecloth or a filter and use quickly while the microorganisms remain healthy.

Dark-looking tea is probably too concentrated. Dilute with water until the color looks like a light iced tea. Test the solution on a few branches to make sure it is an appropriate strength.

Compost tea may be added to flow through irrigation water, but it should not be stored in reservoirs because it will ferment and become unhealthy. The used compost can be spread around plants.

Copper

When copper is dissolved in water, the released ions pass through the cells walls of fungi and attack the proteins inside, killing the fungi. Copper is very effective against gray mold, powdery mildew, *Septoria* leaf spot, and *Pythium*, but it can cause damage to plant tissue. Copper sprays can also be applied as foliar sprays to treat copper deficiencies in plants.

Bordeaux mixture is a combination of hydrated lime (calcium hydroxide [Ca OH$_{2)}$] and copper sulfate (CuSO$_4$) in water that is used in vineyards. It is very effective in controlling gray mold, powdery mildew, and *Septoria* leaf spot. Commercial brands include Dexol Bordeaux Fungicide and Hi-Yield Bordeaux Mix. Bordeaux mixture can be toxic to humans and pets, and use should be reserved for situations of rampant disease outbreak. Use other sprays if possible.

Copper fungicides should be used with care on edible plants. Make sure to follow the directions strictly.

Copper formulas use various compounds that have different toxicity to plants. Before using any copper formula on the whole garden, test it on a portion of a plant to make sure it doesn't cause injury. Apply as little copper as possible and don't apply it in cold, wet weather, which increases copper ion availability and its toxicity to plants.

Some commercially available mixtures containing copper are Kocide, Copper-Count-N, Cueva Fungicide, Concern Copper Soap Fungicide, and Top Cop with Sulfur.

Copper fungicides have been reported to cause toxicological problems in farmworkers in constant contact with them, but they are safe for occasional home use. As with any pesticide, wear protective clothing when working with copper-based formulations.

Coriander Oil

Like some of the other essential oils, such as cinnamon and clove oil, coriander oil has fungicidal and insecticidal qualities. It is effective against diseases such as damping off, *Fusarium* wilt, gray mold, powdery mildew, *Pythium*, and *Septoria* leaf spot, and insects such as aphids, spider mites, thrips, and whiteflies. It is available commercially as a blend in SM-90.

Cultural Controls

Cultural controls are ways of growing plants that prevent or minimize problems with pests or diseases.

Sterilized soil: For indoor or container growing, start out with sterilized soil to prevent the infestation of young plants by pests or pathogens already in the soil.

Air circulation: Space plants to allow for adequate air circulation. Plants grown too close together facilitate a humid environment that may enhance disease incidence, and also provide hiding places for insects.

Avoid overwatering or underwatering: Allow the soil to dry out around plants to control soilborne fungi that grow in conditions of high moisture. Also avoid underwatering, as drought-stressed plants are more susceptible to diseases such as powdery mildew.

Avoid overfertilization: Plants that are over-fertilized, especially with high-nitrogen fertilizers, grow very quickly but are more attractive to pathogens. For general gardening, use fertilizers with balanced nutritional profiles, such as 9-3-6 for leafy growth and 2-10-10 for flowers and vegetables. Always consult the label instructions.

Pruning: Pruning away and destroying diseased leaves or branches of plants removes any disease or pest already inhabiting that tissue. Dispose of any diseased pruned material away from the growing area. Do not compost unless the compost process is sufficiently "hot" to ensure that pests and disease organisms will be killed.

Crop rotation: The rotation of crops susceptible to certain diseases and pests with crops not susceptible denies pests food and places to reproduce, thereby reducing the populations.

Resistant varieties: Planting resistant varieties denies pests and diseases food and places to reproduce. These varieties will tolerate an infestation or disease and yet still produce. Look for VFN varieties. They are resistant to *Verticillium*, *Fusarium*, and nematodes.

Cultivation and seeding: Cultivate the garden space at the end of the growing season. This exposes any overwintering insect larvae or pupae to weathering and predation, and also incorporates green matter into the soil to rot during winter. Plant with a cover crop to cover the exposed soil, stop erosion, and enhance the soil's fertility.

Early watering: Water early in the day. This reduces the water on leaves and around plants so moisture-loving diseases aren't promoted. It also discourages snails and slugs that prefer damp areas.

Avoid weedy growing areas: Weedy areas provide refuge for unwanted garden pests. However, some beneficial insects prefer a weedy border to reproduce or overwinter. Find a balance that works.

Diatomaceous Earth

Diatomaceous earth is made of the hard silica shells of microscopic sea creatures called diatoms. When dry, these sharp, glassy fragments cut into the soft bodies of animals and insects, they die of dehydration. Pests such as aphids, caterpillars, earwigs, leafhoppers, slugs, snails, and thrips can be controlled or repelled with diatomaceous earth, but it also kills beneficial insects and should be used with caution. It is also an excellent product for the control of indoor fleas and lice. A 3-inch-wide barrier deters ants, slugs, and snails from plants and patios. It can also be mixed with boric acid, ground cinnamon, or ground cloves to increase its effectiveness as a deterrent.

Diatomaceous earth is considered non-

Diatomaceous earth can be used both indoors and outdoors but must be dry to be effective. Safer Brand is an organic choice.

toxic to mammals but can be an irritant if inhaled. Wear a dust mask when applying. It is only effective when it is dry, so avoid using it in areas where it will remain wet. Brands include Safer Diatomaceous Earth.

D-limonene

D-limonene is refined from the oils of citrus rinds, has a pleasant citrus odor, and is the active ingredient in many cleaning products. It is made up of two isomers: L-limonene that smells like pines and D-limonene that smells more like oranges. D-limonene is extracted from limonene by using alkali and steam and is the form most often used in cleaning solutions.

D-limonene is a broad-spectrum insecticide effective against ants, aphids, mealybugs, scales, spider mites, and whiteflies. D-limonene also has fungicidal qualities. I've used pure diluted D-limonene and it controlled powdery mildew but did not

eradicate it—perhaps a higher concentration would. It is the active ingredient in Ortho Home Defense Indoor Insect Killer and Concern Citrus Home Pest Control.

ELISA kit

These kits are used to determine the exact disease present in the garden or attacking plants. The term stands for enzyme-linked immunosorbent assay. The technique makes use of an enzyme bonded to a particular antibody or antigen that binds to the target organism or chemistry and indicates its presence by color change. The tests require only a few minutes and are very accurate. Tests are available for the fungi *Botrytis cinera*, *Phytophthora*, *Pythium*, and *Rhizoctonia* as well as for plant pathogenic bacteria such as *R. solani* and for many viruses including tobacco mosaic virus.

The kits are also available to determine the presence of human pathogens such as aspergillis and other phyto-toxins

The kits and test strips' accurate information helps you make informed decisions on how to care for your plant.

Epsom Salts: see Magnesium Sulfate

Fertilizers

Basic garden fertilizers are rated by their NPK numbers. These are noted on the label as a series of three numbers such as 15-5-10, 5-1-1, and so on, describing the content of the three major macronutrients in the product. The first number always represents the nitrogen (N) content of the fertilizer. The number is the percent of the element in the fertilizer. The second represents the equivalent of the phosphorous compound P_2O_5. The third is the equivalent of the potassium (K) compound K_2O (potash). Thus, a fertilizer with NPK values of 10-5-1 contains 10 percent nitrogen, 5 percent equivalent of P_2O_5 phosphate, and 1 percent equivalent of the potassium compound potash. For brevity's sake, they are called N-P-K.

Choose a fertilizer according to the deficiency (if any) that you are trying to treat and also to the growth stage of the plants. Nitrogen promotes early growth and is beneficial to leafy vegetables. Avoid fertilizers high in nitrogen during the flowering stage. Conversely, phosphorous is most needed during flowering and should be used with care during vegetative growth. Potassium is useful at all stages of a plant's growth and is added to balance the pH of high-phosphorous fertilizers.

Fish Emulsion

Fish unsuitable for human consumption, fish bones, and the by-products from fish processing are pressed to remove the oil. The two products remaining after the oil is pressed are a brown powder, fishmeal, and a liquid emulsion, fish emulsion. Both are high-nitrogen fertilizers used for treating nitrogen deficiency and for providing micronutrients. Fish emulsion also has pesticidal and fungicidal qualities. Do not use near living areas because it has a strong stench.

Fish emulsion releases nitrogen quickly to plants, whereas fish meal provides a slower release. Commercially available brands include Alaska Fish Emulsion (5-1-1), Fertrell Liquid Fish Emulsion (4-1-1), Down To Earth Fish Meal (10-4-0), Peaceful Valley Fish Meal (9.6-3-.35), and Dr. Earth Fish Bone Meal (3-16-0).

Fumigation

Sometimes the only way to eliminate gophers is by gassing them. Several "smoke bombs" are sold for this purpose, such as Dexol Gopher Gasser and Revenge Rodent Smoke Bomb. These are thick paper cartridges filled with charcoal, sodium nitrate, and sometimes sulfur. Light the fuse and put it in the gopher's burrow. Toxic fumes from the burning cartridge do the rest.

Garlic

Garlic is antifungal and antibacterial. It is used as an ingredient in pesticides and fungicides and can be prepared as a spray to be used every few days. Garlic's high sulfur content helps to destroy fungi. Garlic also can be added to other antifungal sprays and sprayed on new growth before there is a sign of infection.

Garlic is a general-purpose insecticide as well as a fungicide, so it should be used with caution on outdoor plants. It kills beneficial insects as well as plant pests. Dr. Earth Vegetable Garden Insect Spray, VeggiePharm, Garlic Barrier, and BioRepel are garlic insecticides for ants, aphids, caterpillars, spider mites, thrips, and whiteflies. Garlic fungicides are effective against powdery mildew in brands such as Mildew Cure from SaferGro, Garlic GP Or-

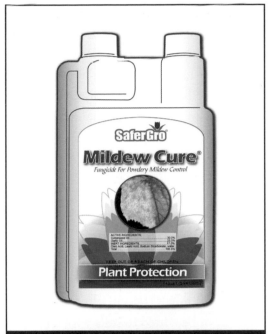

Mildew Cure from SaferGro is made from garlic extract and cottonseed oil. It is an organic fungicide that provides good control of powdery mildew.

namental, and Citrall Lawn and Garden Fungicide. Garlic is also an ingredient in several commercial deer repellents, such as Deer-Off and DeerPharm.

To make a garlic spray:

Mix a teaspoon of pressed garlic juice, available in some grocery stores and garden shops, in a pint of water with 2 ounces of 100 proof or higher drinking alcohol such as rum or vodka. Let sit for an hour or more. Spray.

Granite Dust

Granite dust (or rock dust or stone meal) is the powdery rock leftover from the mining of granite and is a good slow-release source of potassium. Depending on the source, it may also contain a variety of micronutrients that are listed on the label. Brands in-

clude Fishers Creek, Down to Earth, and Agrowinn.

Grapefruit Seed Extract

Grapefruit seed extract is sold as a general-purpose disinfectant, used to control algae in hydroponic systems. Look for brands such as Citricidal and Nutribiotic. Follow label instructions for control of algae, as different brands contain different concentrations.

Greensand

Greensand is a soft, easily crumbled type of sandstone rock composed of glauconite, an iron-potassium silicate mineral, which gives it its dark green color. It is a good source of slow-release potassium and iron. Greensand also provides some phosphorus and small amounts of the micronutrients copper and manganese. Brands include Fertrell Jersey Greensand, Gardener's Supply Company Greensand, and Espoma Organic Traditions Greensand.

Guano

Guano is bat and seabird droppings. Most guanos are high in nitrogen, phosphorus, or both, depending on the source of the guano. Because guanos vary widely, always check the NPK value on the label to make sure that a given guano fertilizer meets your needs. The nutrients are quickly available after it is applied, so it is an excellent way to treat deficiencies or as a general nutri-

Vital Earth makes several guano products that stimulate plant growth.

outdoors to avoid a major odor problem) to dissolve many of the nutrients and make them available for immediate uptake by plants. If you are making a large quantity, you may wish to prepare a concentrate to be diluted later. The resulting solution can be used as a foliar spray or for use in irrigation.

Gypsum

Gypsum is a natural mineral composed of calcium sulfate ($CaSO_4$). It is useful as a slow-release form of calcium or sulfur that doesn't affect the soil pH too much. It should not be added to soils with a pH below 5.5 because it interacts with aluminum (Al) in those acidic soils, making the Al soluble and poisonous to the plants. Gypsum is available at garden centers under various local and store brands.

Gypsum can also be used to break up clay soils.

HEPA filters

These filters are the gold standard of filters because they screen out dust, pathogens, and the tiniest insects. There are several brands of screens designed for this purpose. They include Horti-Control Dust Shroom, Fly-Barr, BugBed 123, and No-Thrips. Even these will only screen out 70 to 80 percent, mostly the adults.

Herbal oils

See Cinnamon, Clove, D-limonene, Garlic, Neem, Oregano, Rosemary, and Thyme.

ent source. Vital Earth has several varieties of guano products, both from seabirds and bats. These fertilizers can increase yields intensely and boost crop yields.

To make the nutrients even more available to plants, simmer the guano in water in a slow cooker for several hours (do this

Herbal oils are Mother Nature's way of controlling pests and disease. These fragrant oils repel and kill. The best time to apply these oils is when the day is cooling down. Herbal oils evaporate over a period of time, leaving no residue. Look for products that blend several oils together for faster, more effective results.

Horticultural Oil

Horticultural oils are any of a number of light oils used to control insects such as aphids, fungus gnats, leaf miners, mealybugs, scales, spider mites, thrips, and whiteflies. They work by smothering insects or by coating the plant with an impenetrable barrier. Horticultural oils are either vegetable or petroleum based. Both are effective, but if you wish to avoid petroleum products, then be sure to check the label. In any case, oils should not be used on soft, edible portions of plants such as lettuce or broccoli buds; they are more appropriate for fruiting plants. These oils need to be distinguished from neem oil, which poisons pests, although, like neem oil, some horticultural oils such as jojoba, sesame, and cottonseed have fungicidal properties. They can be used in combination with other spray ingredients listed here. The oils are mixed at about 1 to 2 percent concentrations. A 1 percent solution is about a teaspoon per pint (5 milliliters per 500 milliliters), 3 tablespoons per gallon (40 milliliters per 4 liters), or one quart in 25 gallons. Add a wetting agent or castile soap to help the ingredients mix.

Horticultural oils are classified as "dormant" oils, which are used on plants during the winter season, and "summer" oils, which are used on growing plants. Summer oils are lighter and more highly refined. Some oils can be used for either purpose. Use only summer oils on annual plants. Dormant oils can harm growing plants. Check the label on any horticultural oil you're thinking of using in your garden to verify that it is rated for use on growing plants. Some suitable brands include Dr. Earth Pro-Active Fruit and Vegetable Insect Spray, Pest Out, Control Solutions Ultra Fine Oil, and Green Light Horticultural Oil.

Hydrogen Peroxide

Hydrogen peroxide (hp, chemical formula H_2O_2) is a contact disinfectant that leaves no residue. It breaks down into water and oxygen. Use it to sterilize instruments such as scissors, trays, tubing, and walls.

Add it to irrigation water to prevent or control algae, gray mold, *Pythium*, and powdery mildew. Hp can be used daily with no adverse effects on the plants. It should not be used if you are fostering microbial life such as mycorrhizae, compost tea, and other root associated organisms.

Hp sold in drugstores has a concentration of 3 percent. Garden shops sell 10 percent hp. ZeroTol contains 27 percent hydrogen dioxide and 5 percent peroxyacetic acid, with an activity equivalent to about 40 percent hp. Both the 10 percent solution and the ZeroTol are hazardous because they burn skin on contact, similar to the burn caused by concentrated acids.

To treat plants with drugstore-grade 3 percent hp, use 4½ tablespoons (70 milliliters) and fill to make a pint (500 milliliters) of solution, or a quart of hp to 2 quarts of water. That makes a 1 percent solution. With horticultural-grade 10 percent hp, use about 4 teaspoons per pint (20 milliliters per 500 milliliters), or 13 ounces per gallon (300 milliliters per 4 liters).

With ZeroTol, use about 2 teaspoons per pint (10 milliliters per 500 milliliters), or 5 tablespoons per gallon (75 milliliters per 4 liters).

Hydroponic Micronutrient Products

Hydroponic fertilizers supply micronutrients in addition to the three macronutrients nitrogen, phosphorous, and potassium. In soil gardens the micronutrient formulas are useful for treating or preventing micronutrient deficiencies. They are available to the plants immediately. There are hundreds of brands available.

Make sure to use products that are balanced for the plant varieties in the garden as well as for their stage of growth to provide the appropriate levels of nitrogen, phosphorous, and potassium. Likewise, if you're trying to treat a specific micronutrient deficiency, make sure that the product you are considering contains large amounts of that nutrient.

Insecticidal Soap

Mild soap solutions damage the exoskeleton of soft-bodied pests and cause dehydration on direct contact with the pests. Aphids, caterpillars, mealybugs, scales, spider mites, thrips, and whiteflies are affected.

Commercial brands include Safer Insect Killing Soap, Dr. Bronner's peppermint castile soap (preferred), Neudorff's Insecticidal Soap, and Monterey Quick. Some soap solutions can damage plants, so test by spraying a branch before applying extensively. Homemade sprays can be made from a few drops of mild soap (like Dr. Bronner's) in a pint of water.

Iron Phosphate

Iron phosphate is one of the best organic means of eliminating slugs and snails. Once they eat a small amount of iron phosphate, the pests stop eating and soon die. It is harmless to plants and pets, unlike poisons such as metaldehyde. Iron phosphate comes in small white pellets under brand names such as Slug-go and Escargo. Sprinkle it liberally around your garden, as well as in shrubbery, ground cover, and other places where slugs and snails like to hide during the day. As it breaks down, it gradually supplies both iron and phosphorous to the garden.

Iron Supplements

These products are used to correct iron deficiencies. Most contain chelated iron, iron sulfate, or iron oxides. Commercial brands include Glorious Gardens Iron Sulfate, Bonide Iron Sulfate, Monterey Dr. Iron, Phyto-Plus Iron 5%, and Biomin Iron.

Kelp Concentrates

Kelp is a family of seaweeds. Some species are harvested and prepared as liquid or granular plant supplements. They contain growth-stimulating enzymes, hormones, and vitamins, as well as macro- and micronutrients. Applying kelp to the garden results in plants with more stamina and more resistance to disease and stress. Kelp is especially useful for protecting seedlings from damping off and for treating potassium and copper deficiencies. Brands include PHC BioPak, Kelp Help Liquid Kelp Concentrate, Gardens Alive Liquid Kelp, Bonide Organic Sea-Green Kelp Extract Concentrate, Maxicrop Seaweed, and Tidal Organics Kelp Meal.

Lime

Lime is a general term for several calcium or calcium-magnesium compounds used to adjust soil ph Up and to correct calcium and magnesium deficiencies. All of these compounds release calcium and sometimes magnesium slowly into the soil.

Hydrated lime—also called slaked lime or calcium hydroxide (Ca [OH]$_2$)—is the most alkaline form of lime suitable for gardening. Only small amounts are needed to adjust the soil pH. It acts very quickly.

Garden lime is calcium carbonate (CaCO$_3$). It is prepared by crushing limestone or oyster shells into powder. It is less alkaline than hydrated lime but will still raise pH. Commercial brands include Planet Natural Oyster Shell Lime and Espoma Organic Traditions Garden Lime. This form adjusts soil gradually.

Dolomitic lime is high in magnesium, making it useful for treating magnesium as well as calcium deficiencies. Brands include Planet Natural Pro-Pell-It and Espoma Organic Traditions Dolomitic Lime, as well as others.

Liquid lime is finely ground garden or dolomitic lime suspended in a liquid. They raise soil pH faster than regular powdered lime. Brands include Turbo Turf Liquid Lime Plus, Aggrene Natural Liquid Lime, and Aggrand Organic Liquid Lime.

Magnesium Sulfate (Epsom Salts)

Also known as Epsom salts, magnesium sulfate (MgSO$_4$) is sold in pharmacies and is the quickest way to correct magnesium and sulfur deficiencies.

A solution of 1 teaspoon Epsom salts per gallon of water may be applied as a foliar spray. To treat planting mixes, use 1 teaspoon per gallon of water, then irrigate. After the initial treatment, treat with 1 teaspoon for 2 to 5 gallons of water if it is not supplied in the fertilizer. Adjust the treatment at first signs of deficiency.

Mechanical Controls

Many pests can be controlled by physically removing and destroying them or by blocking them from reaching their target plants. These methods prevent and minimize infestations. They also supplement other controls. Such methods include:

Air filtration: Fungal spores and small insects can enter an indoor growing space in the airstream, through cooling systems, or through air vents. Fine dust filters placed in the air-intake systems pose a barrier to airstream-borne pests such as aphids, mites, thrips, and whiteflies, and some fungal spores.

Bug zappers: Flying insects such as egg-laying moths are attracted to the blue light and are electrocuted on the charged grid. Unfortunately, beneficial insects like lady beetles are also drawn to the light.

Handpicking: Physically remove and destroy caterpillars, slugs, snails, and other larger pests by hand. Look for slugs and snails in the evening and early morning hours as they hide during the day.

Mulch: A layer of mulch in the garden serves several purposes, including water retention in the soil, a source of decaying plant matter that provides nutrients, and a barrier to insects with soil-dwelling larval stages. The term "mulch" is generally thought of as a layer of finely chipped tree bark but can also be black plastic, gravel, or even sand.

Pheromone traps: These traps come in a variety of shapes, sizes, and purposes, with pheromone baits specific to certain insects. Check labeling carefully to choose the proper trap for the target pest.

Physical barriers: Several types of physical barriers can prevent insects from reaching plants. Vegetable cans or tin foil block cutworms from young plants. Copper wire or tape discourages slugs and snails.

Row covers: Floating row covers are layers of fine cloth that, when draped over crops, prevent insects from reaching the plants. For some crops, however, these layers must be removed to allow for pollination by insects.

UVC Lights: See *UVC Light*

Vacuuming: An ordinary vacuum cleaner can go a long way toward eliminating pests in indoor growing areas such as aphids, caterpillars, and whiteflies..

Water spray: A strong stream of water directed at foliage will knock pests to the ground where they will die or be eaten by predators.

Wiping: Remove mealybugs from plants by wiping them away with a cotton swab moistened with alcohol.

Milk

Milk is an excellent defense against powdery mildew, so much so that rose growers all over the world use it. Use one part milk to nine parts water. The classification of the milk (1 percent, 2 percent, whole milk) varies from recipe to recipe, but 1 percent works well. Some recipes also call for the addition of garlic or cinnamon to the mixture. Use the milk spray at the first sign of infection and spray weekly.

Neem

Neem oil is pressed from the seed of the neem tree (*Azadirachta indica*), a native of Southeast Asia but now cultivated worldwide. Neem is available as an oil, water-soluble liquid, and in dried granules, like Vital Earth's Granular Neem Concentrate. Neem has both insecticidal and fungicidal properties and contains at least 70 different components. Azadirachtin is the most useful. It accounts for 90 percent of the oil's

Vital Earth's Granular Neem Concentrate is a water-soluble powder derived from the neem tree. It leaves no oily residue.

insecticidal activity. Other components, such as nimbin, nimbidin, meliantriol, and salannin, also have fungicidal properties.

Neem has a low mammalian toxicity and is used in toothpaste in India. Once applied, neem degrades quickly, making it safe for the environment, nontarget species, and beneficial insects. It is effective against ants, aphids, bagworms, Colorado potato beetles, cucumber beetles, flea beetles, fungus gnats, lace bugs, leaf miners, mealybugs, Mexican bean beetles, scales, thrips, and whiteflies. It is fungicidal against black spot, gray mold, powdery mildew, *Septoria*, and other insects and diseases.

Neem protects plants from fungi in several ways. First, it has fungicidal properties on contact as it disrupts the organism's metabolism. Second, it forms a barrier between the plant and the invading fungus. Third, it inhibits spore germination. Neem has translinear action—that is, it is absorbed by the leaf and moves around using the leaf's circulatory system. It can also be used as a systemic, which is absorbed and moves throughout the plant.

Neem products are effective as foliar sprays and may also be added to the soil or grow media through irrigation (1 teaspoon neem oil per quart, or 5 milliliters per liter). Then, the neem is taken up by the plant's roots and distributed throughout the plant. As a foliar spray, neem oil should be diluted with water and a dash of wetting agent to a 1 to 2 percent solution. This solution should be used within eight hours. The fungicidal and insecticidal effects are more potent when applied by spray, but the systemic neem lasts longer. For fungicidal applications, neem is best used before the plant or the garden exhibits a major infection. Used in this way, it prevents the fungal spores from germinating.

Azadirachtin and some other components are sometimes purified from the oil with alcohol, leaving behind an oil called clarified hydrophobic extract of neem oil.

From neem oil comes three classes of garden products:

- Broad-spectrum insecticides with azadirachtin such as Vital Earth Neem Granules, AzaMax, Azatrol, Bioneem, and Neemix.

- Fungicides and insecticides that contain hydrophobic extract of neem oil such as Trilogy and Green Light Fruit, Nut, and Vegetable Spray.

- Insecticides containing both azadirachtin and extract of neem oil, or pure pressed and filtered neem oil such as Azatin XL Plus and Dyna-Neem.

Nitrate Salts

Nitrate (NO_3) is the most readily available form of nitrogen for plants. Nitrate salts, such as calcium nitrate ($CaNO_3$) and potassium nitrate (KNO_3), provide a quick shot of soluble nitrogen to treat nitrogen deficiencies. Nitrate salts are available commercially under brand names such as Champion, SQM Hydroponica, and Ultrasol.

Oregano Oil

The main active terpene in oregano is carvacrol, which gives oregano its tangy, zesty, and warming odor, taste, and feeling. It is an arthropodal insect repellant. In addition, it is a fungicide, especially effective against pseudomonas and other plant pathogens. It is also antibacterial, interfering with the organism's cell walls. Use it in plant oil mixes or use as a 1 percent solution with an emulsifier such as lecithin and a wetting agent. Make sure to test on a branch. If it causes leaf damage, dilute with water.

ph Up and pH Down

ph Up and pH down are generic terms for alkaline and acidic pH adjustors, respectively. They are used to change the pH in indoor gardens or aquariums. The active ingredient for ph Up is usually potassium hydroxide (KOH) or potash (K_2CO_3). For pH down, the active ingredient is usually phosphoric acid (H_3PO_4).

Most fungi grow only within a narrow pH range. Raising the pH of leaf surfaces to a pH of 8 makes the environment inhospitable for fungi, such as gray mold or powdery mildew, and prevents or stops growth. Commercial brands include General Hydroponics, GroWell, and Growth Technology.

Improper soil or water pH is sometimes the cause of nutrient deficiencies. Nutrients are soluble or insoluble depending on pH. Insoluble minerals are not available for uptake by roots. Keep soils between 5.8 and 6.3 pH.

Potassium Bicarbonate

Potassium bicarbonate is an effective fungicide against black spot, *Fusarium* wilt, gray mold, powdery mildew, and *Septoria* leaf spot. It raises the pH of the fungi's environment. Available as a wettable powder, potassium bicarbonate is used as a curative as well as a preventative measure, and it also provides potassium to plants with potassium deficiencies.

Most fungi can grow only within a cer-

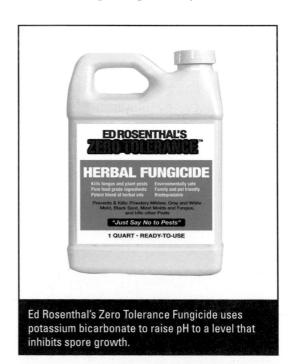

Ed Rosenthal's Zero Tolerance Fungicide uses potassium bicarbonate to raise pH to a level that inhibits spore growth.

tain pH range. An alkaline solution with a pH of 8 makes the environment inhospitable for the fungus and stops its growth. Mixing up such a solution with pH down and applying it as a foliar spray is one of the simplest means of controlling foliar fungi such as gray mold and powdery mildew. It can be used on critically infected plants.

Potassium bicarbonate is more effective as a fungicide when used with an oil and a wetting agent. Use 1 teaspoon (5 milliliters) of potassium bicarbonate, a teaspoon of oil, and a small amount of wetting agent in a pint (500 milliliters) of water, or 3 tablespoons (45 milliliters) each potassium bicarbonate and oil and a half teaspoon of wetting agent in a gallon (3.8 liters) of water.

Use a spray of one-half of 1 percent mixed in water. A wetting agent helps spreads the mix more evenly over the plants.

Other recipes combine potassium bicarbonate with milk or spices such as clove or oregano oil.

Potassium bicarbonate is available in commercial formulations such as Ed Rosenthal's Zero Tolerance Fungicide, Armicarb100, Kaligreen, FirstStep, Remedy, MilStop, and other brands.

Potassium Salts

Potassium salts are highly soluble compounds that can be used to cure or prevent potassium deficiencies without introducing nitrogen or phosphorus. Compounds include potassium sulfate (K_2SO_4)—sometimes called SOP, potassium silicate (K_2SiO_3), and potassium phosphate (K_3PO_4). Potassium salts are available un-

der commercial brands such as Champion, Pro-Tekt, Hydro-Gardens, Planet Natural, and Allganic Potassium.

Predator Urine

Predator urine is one of the two most effective deer repellants. Deer avoid areas that smell of predators such as coyotes. Repellents based on predator urine exploit this behavior. Dog urine and even human urine can also be used, especially if the humans are meat eaters. Stale, strong urine is most effective.

Rather than applying these repellents directly to plants, create a barrier by applying them to the perimeter of the garden. Another method is to string a rope or cord around the garden and tie cloth strips to the cord every 3 feet. Apply the repellent to the cloth strips to deter deer. Look for

Products with multiple modes of action are more effective than single mode products. Safer Brand EndAll blends pyrethrin, neem, and insecticidal soap to kill pests at all stages, from eggs to adults.

repellents with brand names such as CoyotePee and Deer Busters Coyote Urine. Apply the urine weekly so the deer think the predators have visited recently.

Putrescent Eggs

The scent of rotten eggs is one of the two most potent deer repellents. Look for brands such as Deer-Off, Plantskydd, and Liquid Fence. Apply as recommended by the manufacturer.

Pyrethrum

Pyrethrum is a broad-spectrum natural insecticide derived from a plant in the chrysanthemum family. It is a contact insecticide so it must come in contact with target pests. Pyrethrum is toxic to many beneficial fish, insects, and reptiles.

Formulations range from powders to sprays and are available commercially as Safer EndAll, Planet Natural Pyrethrum Powder, and PyGanic Crop Protection.

Quarternary Amines

These compounds are broad-spectrum disinfectants and can be used to disinfect benches, pots, and tools against algae and fungi, including *Fusarium*, gray mold, and *Pythium*. Do not use these compounds directly on plants. Physan20 and Prontech are two commercial brands.

Quarantine New Plants

Always inspect and quarantine new plants for fungal or insect infestations before introducing them into the general population. Keep them in an isolated location or a separate space until you have established that they are pest and disease free, which takes about 10 days. Taking precautions to confirm that plants do not have any hitchhikers aboard reduces the workload on pest management later on.

A good prophylactic practice is to dip new plants and cuttings, even seeds, in a pesticide/fungicide.

Rock Phosphate

Rock phosphate is a naturally occurring phosphate mineral that releases phosphorus slowly and is used as a soil amendment to help prevent deficiencies. "Superphosphate" or "triple phosphate" is not the same material as rock phosphate. Superphosphate and triple phosphate are derived by treating rock phosphate with acids to make the phosphorus more soluble and to concentrate it. Neither of these treated compounds is acceptable in organic gardening.

Rock phosphate is available in garden centers in brands such as Montana Natural, Peace of Mind, and Espoma Organic Traditions.

Rosemary Oil

The active terpenes in rosemary are boreals, camphor, and pinenes. All three of these are insect and mite repellants and killers, with fungicidal and antibacterial qualities. Start with a 1 percent solution.

Rasta Bob's Death Mite is one commercial product featuring rosemary oil.

fective against fungi responsible for damping off, gray mold, powdery mildew, *Pythium*, and *Septoria* leaf spot, and it is insecticidal for mealybugs, whiteflies, and other pests. Commercial brand names include Organocide (a blend of sesame and fish oils) and Green Light Bioganic Home & Garden (a mixture of sesame, soybean, thyme, and wintergreen oils). A 1 percent homemade spray can be made with 1 teaspoon of sesame oil (available at health food stores or gourmet shops) in 1 pint of water and a small amount of lecithin and a wetting agent.

Rotenone

Rotenone is one of the more toxic natural insecticides. It is broad spectrum but is also toxic to beneficial amphibians, fish, insects, and, to a lesser degree, mammals. It kills most chewing insects and is recommended mainly for severe infestations or for insects that have resistance to other insecticides. It breaks down quickly so treated plants are still safe for consumption.

Commercial products containing rotenone (sometimes in combination with pyrethrum) include Bonide Rotenone Dust, Gordon's Garden Guard, and Hi-yield Rotenone 100 Insecticide Dust.

This pesticide should be used only after alternatives have failed.

Sesame Oil

Pressed from sesame seeds, this oil has both insecticidal and fungicidal qualities. It is ef-

Sodium Bicarbonate

See Baking Soda.

Spinosad (*Saccharopolyspora spinosa*)

This naturally occurring, soil-dwelling bacterium produces a powerful insecticidal toxin that is sold under the name Spinosad. The toxin must be ingested to be effective and works well against a large variety of insects. However, it is toxic to bees and should be used with caution around flowering plants. Brands include Captain Jack's Dead Bug Brew, Conserve, and Monterey Spinosad.

Sticky Cards

These cards come in various colors, such as yellow, blue, and pink, and are coated with a sticky substance in which insects become hopelessly ensnared. The color most attractive varies by species. Aphids and fungus gnats prefer yellow while thrips prefer pink and blue. Hang the cards a few inches above the crop, such as with whiteflies, or place cards at container level to capture fungus gnats. Some types of cards also incorporate pheromones.

These cards are used to their best advantage to monitor the insect pest population. They are not very effective as controls.

Sulfur

Sulfur is one of the oldest fungicides and pesticides, used for centuries to control gray mold, and powdery mildew, and *Septoria*. Commercial products containing sulfur are Fertilome Dusting Sulfur and Safer Defender Fungicide. Test-spray a few leaves before applying extensively to plants. In addition to fungicidal qualities, garden sulfur is useful for lowering the pH of alkaline soils and correcting sulfur deficiencies.

Thyme Oil

The main active terpene in thyme oil is thymol, which has insecticidal, antibacterial, and fungicidal qualities. Start by using a 1 percent solution. Terpenes and the plant oils that contain them can be mixed together and often work synergistically. The oils vary in potency between batches so always test a formula before using it.

UVC Light

UVC light is considered deadly to life and kills the spores and tissues of *Botrytis*, powdery mildew, *Pythium* spp., and other fungal pathogens. There are three ways that UVC light is used.

First, there are water sterilizers that kill pathogens in the water supply. As the water passes around the UVC fixture, microorganisms receive fatal rays. The fixtures can also be used to clean inline water in recirculating hydroponic systems after filtration.

Place the fixtures in the ventilation systems of growing rooms to kill fungi and bacteria spores before they enter the growing space. In closed systems, use systems designed to kill pathogens in the air in restaurants. The light is fatal to all airborne organisms passing through it. Some brands of lights are Big Blue, Turbo Twist, and Air Probe Sanitizer.

Handheld UVC wands and automated systems are used to prevent fungal infections. They kill powdery mildew spores on leaf surfaces. The light passes over each plant for only a second each day. This is enough to keep the plants fungus free. Aeon UVC Systems is the only company manufacturing for the garden industry.

Vinegar

Vinegar, like baking soda, is available in almost every kitchen and is toxic to powdery mildew. To make a spray, use 1 tablespoon of vinegar per quart of water. Some gardeners alternate vinegar with potassium bicarbonate.

Adding Dr. Bronner's castile soap to homemade pesticide recipes ensures that the liquid will coat the plant more evenly.

Wetting Agents

Wetting agents are compounds added to solutions to break the surface tension of water, preventing solutions from beading up on plant surfaces and helping them penetrate soil. Commercial brands include Dr. Bronner's peppermint castile soap, Coco Wet, ThermX 70, Phyto Plus Foliar Friend, Natural Wet, and RainGrow Superflow.

Wood Ashes

Wood ashes are useful as a fungicide and a pesticide because they are strongly alkaline. This inhibits growth of fungi. An in-sect that crosses over wood ashes is burned from their high reactivity. Sprinkle wood ashes sparingly around plants but not closer than 2 inches from the plant's stem. It is best not to use wood ashes dry or in solution on plant leaves.

Zinc Phosphide

Zinc phosphide (Zn_3P_2) is a rat poison now seeing a resurgence in popularity. It is fast acting and more specific than some other poisons because it has a strong garlic odor that attracts rodents but repels most other animals. It is sold in baits under brands such as Nu-Kil, Eraze Rodent Pellets, Prozap Zinc Phosphide Oat Bait, and ZP Rodent Bait AG. As with all rat poisons, zinc phosphide baits should be deployed in tamper-proof bait stations in accordance with the manufacturer's instructions. It has secondary toxicity to predatory scavengers, but usually it is ingested in small quantities that are not harmful.

Zinc Salts

Zinc sulfate ($ZnSO_4$) and zinc oxide (ZnO) provide supplemental zinc in cases of zinc deficiency. Brands include Spectrum Chemical, BLU-MIN Liquid Zinc Sulfate, and NutraSul Plus 18% Zinc–Sulfur Fertilizer. Zinc salts are also available in combination with Fe and Mn.

Beneficial Biologicals

Biological controls are purchased as both live organisms as well as biological-based products. They include fungi and bacteria, tiny parasites and predators, both insect and arthropod. Biological controls have advantages because they are harmless to humans and pets and most can be used anytime during the plant cycle. Many reproduce, offering continuing protection. Some sources are Arbico Organics, Tip Top Bio-Control, Koppert, Rincon-Vitova, Nature's Control, and GreenMethods.com.

Ampelomyces quisqualis

This is a beneficial fungus available commercially under the brand name AQ-10. It is effective against powdery mildew.

Aphid Midges
(Aphidoletes aphidimyza)

Aphidoletes aphidimyza is a small predator fly, 2 to 3 millimeters long, that resembles a mosquito. Ferocious consumers of aphids, the larvae are bright orange and are attracted to colonies of aphids because of the honeydew they produce. They are suitable controls indoors and outdoors in gardens, orchards, and trees. The flies overwinter well throughout the country but enter diapause—a period of rest similar to hibernation—when temperatures dip below 40°F, or when the photoperiod is less than 12 hours.

Midges are available as pupae or eggs. Pupae are shipped in vermiculite. Adults emerge from the container, mate, and lay eggs. Larvae hatch and begin to hunt. Several releases, two weeks apart, are most effective. Release a minimum of one per square foot. These beneficials are available from many companies including Nature's Control, Green Methods, Planet Natural, and Hydro Gardens. For home gardens, release 250 midges one or two times, two weeks apart. For greenhouses, release one midge for every one to six plants.

Aphid Midge

Spined assassin bug with prey

Assassin Bugs

Assassin bugs are common predators in outdoor gardens throughout North America but are not available commercially.

Adults are about ¾ inch long, with an elongated body, narrow head, and pronounced snout that curves downward and has three segments. As a member of the Hemiptera family, the bugs have piercing/sucking mouthparts.

Assassin bugs are general predators of a large variety of juvenile and adult insects, especially many caterpillars and maggots such as cabbageworms, Colorado potato beetles, cucumber beetles and cutworms, earwigs, four-lined plant bugs, Japanese beetles, lace bugs, Mexican bean beetles, tobacco budworms, tomato hornworms, and others.

Both adult and nymph assassin bugs capture insects with their front legs covered with bristles, pierce the body of their prey with their proboscis, and suck out their bodily juices. They inject their prey with a paralyzing toxin to make the insect easier to handle and enzymes that liquefy the prey's flesh for easier sipping. They can inflict a painful bite to humans that may cause an allergic reaction and will do so if handled roughly.

Bacillus pumilus

This naturally occurring bacterium produces compounds that kill fungi and inhibits further growth. Strain QST 2808 of this bacterium is patented and used against *Fusarium*, gray mold, and powdery mildew under the names Sonata and Ballad Plus. It is most effective when applied as a preventative measure. It is often used with *Bacillus subtilis* for more complete control.

Another strain, Strain GB34, is marketed as Yield Shield Concentration Biological Fungicide for the control of various root rots, including *Fusarium* and *Rhizoctonia*.

Bacillus pumilus is marketed as Sonata and Custom B5 by Custom Bio.

Bacillus subtilis

Bacillus subtilis works against various fungi, and several strains of it are patented for use against black spot, *Fusarium*, gray mold, powdery mildew, *Pythium*, and *Verticillium*. It is most useful as a preventative measure and should be applied before symptoms appear, but it can also be used to control

Serenade is a *Bacillus subtilis* product used for powdery mildew, black spot, and other plant infections.

infections. For more effective control, use it with *Bacillus pumilis*.

Bacillus subtilis is available commercially as Serenade, Cease, Custom B5, Rhapsody, and Kodiak.

Bacillus thuringiensis (Bt)

Bacillus thuringiensis, or Bt, is a naturally occurring bacterium that is lethal to insects. Several different strains of the bacterium are marketed to target specific pests, so read the label carefully to match the product to the pest. After eating or touching the bacterium, the host stops feeding and dies within a few days. The bacterium emerges, ready to infect the next victim. This is insect plague, but the bacteria are not interested in you unless you are an insect. Bt establishes itself in the garden and provides protection against infestation or reinfestation.

One species, *Bacillus thuringiensis*, var. *kurstaki* (Bt-k), targets caterpillar, moth, and butterfly larvae in the order Lepidoptera. Brands include Garden Dust and Caterpillar Killer from Safer Brand, Javelin, and Thuride.

Bacillus thuringiensis, var. *israelensis* (Bt-i), is effective against fungus gnats. It is marketed commercially under the brand names VectoBac and Gnatrol.

Beauveria bassiana

This common fungus controls a variety of insects and other arthropods with a soil-dwelling life stage, including aphids, cabbageworms, Colorado potato beetles, European corn borers, grasshoppers, Japa-nese beetles, leaf miners, spider mites, and whiteflies. Unfortunately, many beneficial insects such as lady beetles are also susceptible. Products containing this fungus work on contact and take three to seven days to kill the pest. Spray plants thoroughly as soon as an infestation is evident. *B. bassiana* also works as a preventative against infestation and reinfestation.

Products containing *Beauveria bassiana* include Mycotrol and Botanigard.

Big-Eyed Bugs (Geocoris spp.)

Big-eyed bugs are insect predators found all across the United States and are common in most gardens, but are not available commercially.

They are small, fast-moving insects, generally measuring from ⅛ to ¼ inch long. Their bodies are oblong, usually gray, brown, or black, with a broad head and

Big-Eyed Bug

large, wide-set eyes on each side. Wings are clear and visibly overlap over the back. Most have tiny spots on their heads and thoraxes. Nymphs are similar in appearance to the adults but are wingless.

Prey include aphids, cabbage loopers, caterpillars, chinch bugs, flea beetles, leafhoppers, plant bugs, Mexican bean beetles, spider mites, thrips, whiteflies, the eggs of the corn earworm, and some bollworms. Both adults and nymphs catch prey, pierce the bodies with their piercing/sucking mouthparts, and suck out the liquefied flesh.

They thrive in uncultivated areas around the garden with clover, cosmos, goldenrod, common knotweed, pigweed, and soybeans.

Adult big-eyed bugs are voracious. They consume several dozen spider mites per day.

Damsel Bugs

Damsel Bugs *(Nabis spp.)*

Damsel bugs are found throughout North America and are common visitors in gardens but are not available commercially.

Adults are long and slender with mantis-like front legs used for grasping prey. They hold their front legs up, as if delicately lifting a skirt—hence the name "damsel bug." Most are about ¼ inch long, colored brown, gray, or tan, and they have curved beaks and bulging eyes on a narrow head. Nymphs resemble adults except that they are smaller and have no wings.

Adults and nymphs prey on aphids, asparagus beetles, caterpillars, Colorado potato beetle eggs and larvae, four-lined plant bugs, leafhoppers, sawfly larvae, spider mites, thrips, whiteflies, and other insect eggs. Damsel bugs have piercing/sucking mouthparts. They catch their prey with their slender front legs, pierce their bodies with their beaks, and suck out the contents.

They are attracted to low-growing, evergreen perennials such as alfalfa, candytuft, clover, creeping phlox, dianthus, lavender, lilyturf, and thyme.

Damsel bugs can exist for up to two weeks without feeding, but they turn to cannibalism when other food is not available.

Gliocladium virens (beneficial fungus)

Gliocladium virens is a genus of beneficial, soil-dwelling fungi that destroy pathogenic fungi responsible for damping off and root rot. It works best as a preventative, applied as a soil drench before any symptoms of infection appear. Fungicides containing *Gliocladium* include Gliomix, SoilGard, Primastop, and Prestop.

Ground Beetles

There are thousands of species of ground beetles. They are found in all ecological niches all over the world including your garden. They are not available commercially. Larvae are grubs, dark brown or black, with 10 body segments that narrow toward the end of the body. The adults are shiny and black, blue, gold, metallic green, or red, with longitudinally ridged wing covers, a head smaller than the thorax, and threadlike antennae. They vary from ⅛ to 1½ inches in length. Both grubs and adults have large mandibles built for capturing and devouring their prey. They are fast moving and scurry away when disturbed.

Ground beetles prey on beetles, caterpillars, cutworms, grasshoppers, gypsy moth larvae, maggots, slugs, snails, and other insects that have a ground-dwelling larval stage. They capture prey and consume them with their powerful jaws.

Ground beetles usually hunt at night and hide during the day. While they often wander into homes by crawling un-derneath the door or into crevices, they are not household pests and prefer to be outdoors. They overwinter as adults under debris, around rotting wood, or in the soil.

Place boards, rocks, or stepping-stones in the garden to provide hiding places.

Provide permanent plantings of clover and other perennial plants around the garden.

Ground Beetle

Hoverflies (Syrphid Flies, Flower Flies)

Hoverflies are common across North America and are commonly found in gardens, but they are not available commercially. Adults are yellow, white, or orange with black stripes, about ½ to ⅝ inch long. They earn their name by hovering over flowers to sip the nectar, much like hummingbirds. Despite their similarity in appearance to bees, they do not sting and are not harmful to humans. In fact, the adults are primarily pollinators and it is the larvae that do the pest control.

Larvae are slug-like maggots, gray or greenish in color. A dark, sticky excrement is evidence that the larvae are present. Eggs are white and cylindrical, laid singly in areas with plenty of aphids. Hoverflies prefer aphids, but they also prey on cabbageworms, caterpillars, mealybugs, scales, and thrips.

The larvae crawl along plant surfaces searching for prey. Once they find a likely victim, they seize the insect, suck out its contents, then discard the skin. One larva can consume hundreds of insects in a month. High populations of hoverflies can

Flower Flies

Flower fly larva

control from 70 to 100 percent of an aphid infestation.

Pupae overwinter in the soil or in debris and emerge in late May or June. Female flies lay their eggs in areas with high aphid populations. The eggs hatch in two or three days and the young larvae begin to hunt, feeding for three or four weeks before pupating. Adults emerge in about two weeks. Depending on the species, there may be two to seven generations per year.

Attract them with a variety of nectar- and pollen-producing plants such as *Alyssum*, aster, calendula, coreopsis, cornflower, cosmos, daisies, dill, fennel, feverfew, lavender, marigolds, mint, statice, sunflowers, wild mustard, and zinnia. Time plantings so that something is in bloom throughout the gardening season to provide food for the adult flies.

Many species of hoverflies mimic bees, wasps, and yellow jackets in both appearance and behavior to protect themselves from predators. One species waves its front legs in front of its face to mimic the waving antennae of the potter's wasp. These guises are thought to be one way the flies protect themselves from predators. But a closer look at their wings reveals that bees and wasps have four wings and flies have two.

Lacewings *(Chrysopa spp.)*

Lacewings are found across North America and are common garden visitors. They are also available commercially. The larvae are ferocious predators of aphids, small caterpillars, whiteflies, and any other small insects they can catch. The adults may eat

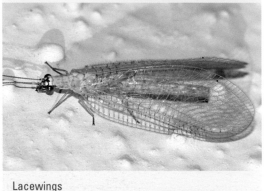

Lacewings

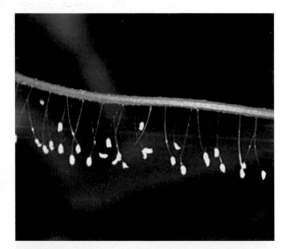

Lacewing eggs

lunges forward and impales the prey with its hollow mandibles. It injects digestive juices into the victim to liquefy its insides and then sucks this slurry out.

Adult lacewings are light green or brown, fragile-looking insects with large, golden eyes and threadlike antennae. Their huge, intricately veined wings, for which they are named, are held over their backs when not in flight. Species measure from ½ inch to 1 inch in length. Larvae are fast moving, up to ½ inch long with a flattened body with large mandibles for catching and holding prey so they look like caterpillars with an alligator-like mouth. Green lacewings attach single eggs on filamentous stalks to protect them from their cannibalistic siblings. Brown lacewing females glue their eggs to buds or twigs.

Use lacewings in greenhouses but not indoor gardens, where they dive into lights. Outdoors, the adults tend to fly away once they emerge from the larval stage, so they are most effective in large gardens and fields. Adults are attracted to gardens with plants that are copious pollen producers such as angelica, *Coreopsis*, cosmos, daisies, dandelions, goldenrod, Queen Anne's lace, sunflowers, tansy, and yarrow. They are top feeders so they fly seeking better sources when food becomes scarce. At night they are attracted to light and can be seen around lights and clinging to lit screened windows.

Two species are available commercially: *Chrysopa carnea* and *Chrysopa rufilabris*. Buy them as either larvae or eggs from companies including Bio Ag, Gardens Alive, Rincon-Vitova, Planet Natural, Natural Insect Control, and American Insectaries. Use about 1,000 eggs or 200

some aphids but live mostly on pollen. The main drawback to lacewings is that you can't use them with other beneficial insects: the lacewings and their larvae are such fierce predators that will eat other beneficials, or even each other, if food is scarce.

Lacewing larvae patrol plants for prey and detect it via contact. Then the larva larvae for every 100 square feet (9 square meters). Spread the larvae out upon release because they are cannibalistic. Time releases two weeks apart.

Lady Beetles (Ladybugs)

The lady beetle (also called ladybug or ladybird beetle) is common throughout North America and is ubiquitous. There are over 450 species on the continent. They are also available commercially.

Larvae look a bit like alligators with a large, sickle-shaped mandibular jaw and short spiny projections along their bodies. They are blue black with orange spots and six legs.

The adults are round with domed backs, about ¼ inch long, reddish to yellow in color, and with or without prominent spots on their backs. The head and the legs are black. Females lay yellow, oval eggs, end up, in clusters of 10 to 50 on the undersides of leaves.

Both lady beetle adults and larvae are voracious general pest predators of aphids, beetles, caterpillars, lace bugs, mealybugs, mites, scale, whiteflies, and insect eggs.

Lady Beetles

"The Argument" between ant and lady beetle

Many predator insects including lady beetles, lacewings, and hoverflies require a partially vegetarian diet. Good Bug Blend from Arbico satisfies this need and keeps them reproductive.

The larvae are the more voracious of the two, able to consume up to 40 aphids per hour—their favorite food. Adults do serious damage to pest populations as well. A female lady beetle consumes 300 aphids before she starts to lay eggs, and as many as 5,000 aphids over her lifetime.

Lady beetles are best suited to greenhouse and outdoor gardens because they make suicide dives at hot lights indoors. They are top feeders and fly away when food levels drop so if you have a small garden, it's likely that purchased adults will exit from it shortly after being placed there.

Adults feed on pollen so a wide range of flowering plants helps attract and keep them on-site.

Choose the species depending on your pest. *Hippodamia convergens* attacks aphids, while *Cryptolaemus montouzieri* and *Ryzobius lophanthae* devour mealybugs and scale. They are widely available from insectaries, including Arbico Organics, Bio Ag, Gardens Alive, Nature's Control, Rincon-Vitova, and EcoSolutions. Use about 150 adults or larvae for every 100 square feet (9 square meters).

Mantids (Praying Mantids)

Mantids are one of the most recognizable beneficial insects in the garden. About 20 native species live throughout North America, along with other species introduced for pest control.

Mantids are indiscriminate hunters that feed on a wide variety of insects, including aphids, beetles, caterpillars, crickets, earwigs, flies, four-lined plant bugs, grasshoppers, leafhoppers, Mexican bean beetles, moths, and squash bugs. They also eat beneficials such as bees and butterflies and are cannibalistic.

Mantids are very efficient and deadly ambush predators, waiting for their victims rather than hunting. When prey is sighted, they grab it with their front legs. Tiny spines on the inside of their front legs help hold the prey in place while it is eaten alive. Mantids are territorial, and only one adult is usually found on a plant.

When mating, the male mantid leaps onto the back of the female. Sometimes the female turns her head and eats the male's head but its body is still able to complete the act. The female may consume the rest of the male's body as an after-sex treat.

The eggs overwinter in rectangular egg cases that the insect hangs from sturdy branches. The case skin is gray and hardens to a Styrofoam-like consistency. The eggs hatch in spring and nymphs disperse and immediately begin to hunt. When released, they roam looking for their own territory so only a few are likely to remain in a small garden. They are more appropriate for large spaces.

Usually, only one generation per year is produced.

The adults of these insects are up to 4 inches in length. They are either brown or green, depending on the species, so they blend into the environment. Females have a heavier abdomen than the males. Their front legs are enlarged for catching and holding prey and are usually held in a "praying" position. Prominent eyes on either side of the head can swivel 180 degrees. Adults have wings but usually do not fly unless searching for a mate. Nymphs are very similar to the adults except they are smaller and do not have wings.

Mantids are available widely from nursery suppliers and gardening catalogs. The Chinese mantid (*Tenodera aridifolia*

Mantid

sinensis) is the one most commonly sold species. The Chinese mantid is not as winter hardy as the native species but is a more voracious feeder.

Minute Pirate Bugs *(Orius spp.)*

The minute pirate bug (*Orius insidius*) is a serious predator; it uses its piercing needle-like beak to suck the life out of its victims. Minute pirate bugs are indigenous to North America and are commonly found in gardens and are available commercially. Both adults and nymphs prey on aphids, beetles, caterpillars, insect eggs, lacebugs, leafhopper nymphs, scale, spider mites, thrips, tobacco hornworms, and whiteflies.

These quick-moving insects are small, about ⅛ to ¼ inch long. Adults are oval and black-and-white patterned. Nymphs are teardrop shaped, shiny, and wingless. They begin life yellow, change to orange, and then to mahogany brown as they mature.

Adults overwinter in the crevices of bark, in weeds, and in plant debris. In spring, they emerge and females lay eggs inside plant tissue. Eggs hatch in three to five days, and the nymphs feed for two to three weeks before becoming adults. Two to four generations are possible per year. They are among the first predators to emerge in spring and rely on pollen and nectar to sustain them until prey is available. Also, when pest populations drop off, pollen and nectar from alfalfa, daisies, goldenrod, oregano, parsley, sage, stinging nettle, wallflower, wild mustard, and yarrow again become a valuable food source.

Minute Pirate Bug

Minute pirate bugs are available from insectaries shipped as adults in bran, rice hulls, or other inert material along with a temporary food source. Release them at the rate of one pirate bug per square foot (10 per M². They are available from Planet Natural, Rincon-Vitova, Bioplanet, and Natural Insect Control.

Minute pirate bugs inflict a painful bite that doesn't pierce the skin.

Mites, Predatory

Predatory mites are predaceous on other species of mites, such as spider mites. Once introduced, this beneficial insect breeds quickly. Some mites, such as *Hypoaspis miles*, live in the soil, attacking root pests, and others live on the foliage. Mites such as *Metaseiulus occidentalis* are used to control spider mites on fruit trees and *Phytoseiulus persimilis* to control spider mites in large-scale plantings of flowers, strawberries, and vegetables. Predatory mites also eat fungus gnats and thrips. Look for species such *Amblyseius, Galendromus, Hypoaspis, Neoseiulus*, and *Phytoseiulus*.

Which Mite Is Right?

PEST	PREDATORY MITE
Aphids	*Iphiseius (Amblyseius) degenerans* *Neoseiulus (Amblyseius) californicus* *Phytoseiid mites*
Burrowing larvae and pupae	*Hypoaspis miles*
Caterpillars	*Iphiseius (Amblyseius) degenerans*
Fungus gnats	*Hypoaspis miles*
Leafhoppers	*Neoseiulus (Amblyseius) californicus*
Mealybugs	*Neoseiulus (Amblyseius) californicus*
Root Aphids	*Hypoaspis miles*
Spider mites	*Galendromus occidentalis,* *Iphiseius (Amblyseius) degenerans* *Mesoseiulus longipes* *Neoseiulus (Amblyseius) barkeri* *Neoseiulus (Amblyseius) californicus* *Neoseiulus (Amblyseius) cucumeris* *Neoseiulus fallacis (Amblyseius fallacis)* *Phytoseiid mites* *Phytoseiulus persimilis*
Thrips	*Hypoaspis miles* *Iphiseius (Amblyseius) degenerans* *Neoseiulus (Amblyseius) barkeri* *Neoseiulus (Amblyseius) californicus* *Neoseiulus (Amblyseius) cucumeris* *Phytoseiid mites*
Whitefly eggs	*Iphiseius (Amblyseius) degenerans*

Apply predators more heavily when pest densities are high and fewer predators when they are less dense. If food is scarce where the mites are located, they migrate. While predatory mites can be introduced in high concentrations to control an epidemic, the best strategy is to lower the pest population using other controls first. This reduces stress on the plants. Then add the predators to control the residual population.

Gardeners can buy them as adults from companies such as Natures Control, Rincon-Vitova Biocontrol Network, Buglogical, and Planet Natural. Plan on using about 30 to 50 adults per plant, or 300 per 100 square feet (9 square meters). Note that while predator mites do reproduce quickly, they can't play catch-up with a rampant spider mite infestation. If your infestation is already out of control, then

Amblyseius

use a nonpersistent control to reduce the spider mite population before releasing predator mites.

Predatory mites are living organisms. To maximize their efficacy, use them as soon as you obtain them. Place them directly on plant leaves no matter what carrier they are shipped on (usually bean leaves or vermiculite).

Each predator species has its own environmental requirements and preferred diet.

Mycorrhizae

These beneficial soil fungi live a symbiotic life with plant roots. Their name, mycorrhizae, literally means "fungus roots." The name defines their role and relationship. The fungi colonize plant roots and extend deep into the soil, becoming true extensions of the roots themselves by increasing the water and nutrient-absorption areas of the plant roots by 100 to 1,000 times.

In addition to improving the water-uptake abilities of plants, the fungi release powerful enzymes into the soil that dissolve hard-to-obtain nutrients such as organic nitrogen, iron, and phosphorus. There are two types of these beneficial fungi: fungi that wrap around the plant's roots forming sheaths at the root tip (ectomycorrhizae) and those that actually penetrate the plant's root cells (endomycorrhizae).

Benefits include:

➤ Improved nutrient and water uptake

➤ Improved root growth

➤ Improved plant growth and yield

➤ Reduced transplant shock

➤ Reduced drought stress

Commercial products containing mycorrhizae include Vital Earth Mycorrhizal, MycoMinerals, MycoGrow, Earth Juice Rooters Mycorrhizae, Plant Success, and SoilMoist.

Nematodes, Predatory

Predatory nematodes prey on the soil-dwelling larval stage of many insects such as cucumber beetles, cutworms, earwigs, flea beetles, fungus gnats, Japanese beetles, squash vine borers, and thrips. Two spe-

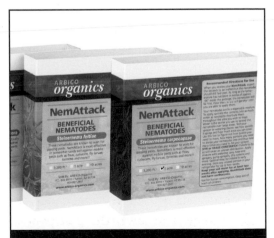

Nematodes attack insects and their larvae and pupae in the soil. Arbico offers a choice of nematodes as well as a combo pack that provides comprehensive control.

cies that are commonly used, *Steinernema feltae* and *Heterorhabditis bacteriophora,* live for months in the soil and can be used as both a curative and a preventative measure. They are harmless to plant roots, and they are not sensitive to most pesticides. These species die from dry conditions, exposure to UV light such as sunlight, and storage in hot conditions. Nematodes are most active when the soil or media is between 68°F and 85°F.

Predatory nematodes usually come as spray solutions or in sponges that are soaked in water; then the water is applied as a spray or soil drench. These can be purchased from Arbico, Orcon, Scanmask, and NemaShield.

Apply in the evening or on cool, cloudy days to allow the nematodes time to burrow into the soil and avoid desiccation. Reapply at six-week intervals.

Watering from the top of containers flushes nematodes from the top of the soil, making the treatment ineffective. Use sub-irrigation, watering from the bottom.

Some beneficial nematodes include *Heterorhabditis bacteriophora,* for beetles and thrips. Brands include NemaSeek, Heteromask, and Terranem. *Steinernema feltiae* is an option for fungus gnats and thrips. Brands include NemAttack, Ne-

maShield, Scanmask, and Entonem. Use *Steinernema carpocapsae* for caterpillars, fungus gnats, and some larvae. *S. carpocapsae* dwells near the soil surface, about 2 inches (5 cm) deep.

Pseudomonas capacia

Pseudomonas is a genus of beneficial bacteria that are fungicidal. Several different species specifically control *Pythium, Fusarium,* gray mold, and damping off fungi. Products containing these bacteria include BlightBan A506 (for apples and pears) and Bac-Pack. Different species of *Pseudomonas* control different pathogens, so check the label to make sure the product matches the targeted pest.

The bacteria should not be used around people with compromised immune systems. It has been identified as a casual agent in infections of cystic fibrosis patients.

Robber Flies

There are over 1,000 species of robber flies in the United States, and they are common in gardens.

Adult robber flies are large; some species grow to 1½ inches in length. They are generally gray to black with segments that

Robber Flies

ny spot. When prey is sighted, they zoom off, nab their prey in midair, inject digestive enzymes to liquefy its insides, and move to a shady spot to enjoy their meal, sucking out the resulting slurry. Then robber flies move back to their sunny perch.

In fall females lay eggs near the ground. The eggs hatch, and the robber fly larvae live beneath the surface of the soil for one to three years. In spring, they resume feeding on decaying plant matter and soil-living larvae of other insects. They pupate early in summer, emerging as adults in fall.

Streptomyces griseoviridis

This beneficial bacterium produces fungicidal agents effective against *Fusarium*, gray mold, and *Pythium*. It is available as Mycostop.

Streptomyces lydicus

This bacterium is effective against stem rot diseases *Fusarium*, *Pythium*, *Phytophthora*, *Rhizoctonia*, and *Verticillium*. It also attacks foliar diseases like black spot, gray mold (*Botrytis*), powdery mildew, and rusts.

Streptomyces lydicus is a growth-enhancing bacterium with antifungal properties. It colonizes roots, reducing the area that pathogens can attack. It produces fungicides detrimental to fungal pathogens. One of these enzymes, chitinase, breaks the fungal cell wall, and the bacteria digest the nutrients of this organism. It gains entry into the fungal hypha by emitting enzymes that break down cell walls. It is available as Actinovate.

may be patterned, banded with a contrasting color. They are hairy. Their bodies are elongated, tapering toward the end. Their eyes are large and prominent with a divot or depression on the top of their head between them. Their legs are long, strong, and bristled to capture and hold prey. Larvae live beneath the soil and are tiny, slightly flattened maggots.

Adults and larvae prey on bees, Colorado potato beetles, dragonflies, flies, four-lined plant bugs, grasshoppers, Japanese beetles, Mexican bean beetles, spiders, and wasps. Adult robber flies perch on a specifically chosen location, usually an open, sun-

Tachinid Fliy

Tachinid Flies

Tachinids are a large and important family of parasitic flies found throughout North America. More than 1,300 species are represented, including many imported for pest control. Tachinid flies are not available commercially.

Tachinid fly larvae parasitize armyworms, cabbage loopers, Colorado potato beetles, corn earworms, cucumber beetles, cutworms earwigs, four-lined plant bugs, gypsy moth larvae, Japanese beetles, Mexican bean beetles, sawfly larvae, squash bugs, stink bugs, tent caterpillars, and tobacco budworms. Some species are specific, seeking one particular host.

Adult tachinid flies resemble the more recognizable housefly except that they are more robust with stout, bristly hairs on the abdomen. They are usually dark gray, black, or brown, and about ⅓ to ¾ inch long. Larvae are maggots that are found inside the parasitized host's body.

The female tachinid fly lays an egg either in, on, or near a host. When the egg hatches, the larva consumes its host's non-

vital tissue first. Some other species lay their eggs directly onto foliage that host species then eat, ingesting the larva that feeds inside the host's digestive system. This last method is particularly true of caterpillars. In some species, the female actually places a live larva onto a host and the larva then burrows in.

As the tachinid fly larva feeds, it slowly kills its host. Once mature, it pupates either inside the host or else drops to the ground and pupates in the soil. Two generations per year are possible. Adult tachinid flies feed on nectar and pollen and are important pollinators as well as excellent predators. They are often seen sitting on foliage

Trichoderma

Several species of this genus of beneficial fungi prevent infestations by gray mold, *Fusarium, Pythium, Septoria,* and *Verticillium.* Some commercial products such as PlantShield and RootShield use a particularly effective strain of the species *Trichoderma harzianum,* KRL-AG2, that was developed at Cornell University.

Wasps, Parasitoid (Chalcid, Aphidius [braconid]), *Encarsia formosa,* Ichneumon, and *Trichogramma* wasps)

Several species of wasps act as parasitoids on garden pests. These wasps are nonsocial, stingless, and so tiny that once you release them you may never see them again. The

Braconid wasp parasitizing a gypsy moth caterpillar

"Mummies"—Braconid wasp young hatch inside the aphid and eat their way out.

entire life cycle of parasitoid wasps revolves around the host pests. They lay their eggs in eggs, larvae, or adult pests (depending on species); then wasp larvae consume the pest from within. The adult wasps of some species eat pests, too.

The victim dies in seven to 10 days after the wasp egg is injected, and the mature wasp pops out of the host's mummified shell.

Indoors and in small outdoor gardens, use about 1 pupae per square foot, 10 per square meter. Some pests require multiple applications for full control indoors and in small outdoor gardens. Two releases, about a week apart, are more effective than a single release.

Different wasp species use different insect hosts, as shown in the following chart.

Buy parasitoid wasps as pupae from companies such as Arbico, Buglogical, Hydro-Gardens, IPM Labs, Nature's Control, Planet Natural, and Rincon-Vitova.

Which Wasp Is Right?

PEST	PARASITOID WASP
Aphids	*Aphidius* and *Aphelinus*
Caterpillars	*Trichogramma*
Fungus gnats	*Synacra pauperi*
Leafminers	*Dacnusa*, *Diglyphus*, and *Opius*
Mealybugs and scale	*Metaphycus*, *Leptomastiz*, and *Anagyrus*
Whiteflies	*Encarsia*
Cotton aphid, melon aphid, green peach	*Aphidius colemani*
Green peach aphid	*Aphidius matricariae*
Potato aphid	*Aphidius ervi*

PARTIAL BIBLIOGRAPHY

Adam, Katherine L. "Squash Bug and Squash Vine Borer: Organic Controls." National Sustainable Agriculture Information Service. 2006. http://attra.ncat.org/attra-pub/squash_pest.html (accessed March 2011).

Agrios, George N. Plant Pathology. New York: Academic Press, 1969.

Bachmann's. Plants for Acid and Alkaline Soils. 1997. http://www.bachmans.com/Garden-Care/divHomePage.html?cnb=GardenCare&categoryCode=02&pageIndex=_pageIndexToken_plantsforAcidAlkalineSoils.

Bader, Dr. Myles H. 1001 All-Nautral Secrets to a Pest-Free Property. Dr. Myles H. Bader. 2005.

Bailey, Stephanie. "Ground Beetles." University of Kentucky: College of Agriculture. 1993. http://www.ca.uky.edu/entomology/entfacts/ef104.asp.

Baker, Jerry. Old-Time Gardening Wisdom. Wixom, Mich.: Jerry Baker. 1999.

Ball, Jeff. Rodale's Garden Problem Solver: Vegetables, Fruits, and Herbs. Emmaus, Penn.: Rodale Press. 1988.

Beckerman, Janna and B. Rosie Lerner. "Purdue University." Salt Damage in Landscape Plants. 2009. http://docs.google.com/viewer?a=v&q=cache:Ysj0FR-VFCAJ:www.extension.purdue.edu/extmedia/ID/ID-412-W.pdf+salt+damage+plants&hl=en&gl=us&pid=bl&srcid=ADGEESgu3sRbCEwI-ISaft6jLNL_IBs1wDeUp_sO0cC3V-M7d942WjbOjtrH8eEWFKPBCrXf_pUFAiPRg5RrR499XM3p4ppUyJ9NvRUpO.

Berry, Ralph E. "Western Damsel Bug." In Insects and Mites of Economic Importance in the Northwest, 221. 1998.

Berry, Wade. "Symptoms of Deficiency In Essential Minerals." Plant Physiology Fourth Edition Online . 2004. http://4e.plantphys.net/article.php?ch=3&id=289.

Bessin, Ric. "University of Kentucky: College of Agriculture." Squash Vine Borer. http://www.ca.uky.edu/entomology/entfacts/ef314.asp (accessed March 2011).

Bolda, Mark. "Strawberries and Caneberries." Agriculture and Natural Resources, University of California. 2011. http://ucanr.org/blogs/blogcore/postdetail.cfm?postnum=1874.

Bradley, Fern Marshall, Barbara W. Ellis and Deborah L. Martin. The Organic Gardener's Handbook of Natural Pest and Disease Control. New York, NY: Rodale , 2009.

Brown, Patrick H. and Barry J. Shelp. "Boron mobility in plants." Plant and Soil, 1997: 85-101.

Burkness, Suzanne and Jeffrey Hahn. "Squash Vine Borer Management in Home Gardens." University of Minnesota Extension. 2011. http://www.extension.umn.edu/distribution/horticulture/M1209.html (accessed March 2011).

Buss, Eileen A. "Leafminers on Ornamental Plants." University of Florida IFAS Extension. June 2006. http://edis.ifas.ufl.edu/mg006 (accessed March 2011).

Capinera, John L. "University of Florida Institute of Food and Agricultural Sciences." Featured Creatures: Imported Cabbageworm. September 2008. http://entnemdept.ufl.edu/creatures/veg/leaf/imported_cabbageworm.htm (accessed March 2011).

Comer, Gary L. and Amanda D. Rodewald. "Ohio State University Extension Fact Sheet: Effective Mole Control." The Ohio State University. http://ohioline.osu.edu/w-fact/0011.html.

Cornell University. "Plant Disease Diagnostic Cllinic Fact Sheet: Botrytis Blight." Cornell University . 2009. http://plantclinic.cornell.edu/factsheets/botrytis/botrytis_blight.htm.

Cornell University. "Powdery Mildew." Plant Disease Diagnostic Clinic. 2009. http://plantclinic.cornell.edu/FactSheets/powdery/powdery.htm.

Cranshaw, W. and R. Foster. "Perdue: Alternative Control Outreach Research Network." Imported Cabbageworm. http://www.agriculture.purdue.edu/acorn/pest.aspx?pest=Imported_Cabbageworm (accessed 2011).

Cranshaw, W. S. "Hornworms and "Hummingbird" Moths." Colorado State University Extension. February 2007. http://www.ext.colostate.edu/PUBS/INSECT/05517.html (accessed March 2011).

Cranshaw, W. S. and R. A. Cloyd. "Colorado State University Extension." Fungus Gnats as Houseplant and Indoor Pests. 2009. http://www.ext.colostate.edu/pubs/insect/05584.html.

Cranshaw, W.S., D.A. Leatherman and J.R. Feucht. "Leafmining Insects." Colorado State University Extension. January 2009. http://www.ext.colostate.edu/pubs/insect/05548.html (accessed March 2011).

Cranshaw, Whitney. Garden Insects of North America. Princeton, New Jersey: Princeton University Press, 2004.

Devlin, Robert M. Plant Physiology. Reinhold Publishing Corporation, 1966.

Diver, Steve. "Cucumber Beetles: Organic and Biorational Integrated Pest Management." National Sustainable Agriculture Information Service. 2008. http://attra.ncat.org/attra-pub/cucumberbeetle.html#species (accessed February 2011).

Dreistadt, S. H. and E. J. Perry. "Pests in Gardens and Landscapes: Lace Bugs." University of California Agriculture and Natural Resources. August 2006.

Dreistadt, S. H. and E. J. Perry. "Pests in Gardens and Landscapes: Lace Bugs." University of California Agriculture and Natural Resources. August 2006. http://www.ipm.ucdavis.edu/PMG/PESTNOTES/pn7428.html (accessed March 2011).

Ecochem: An Earth Friendly Company. Boron. 2011. http://www.ecochem.com/t_micronutrients.html.

Ecochem: An Earth Friendly Company. Copper. 2011. http://www.ecochem.com/t_micronutrients.html.

Ecochem: An Earth Friendly Company. Manganese. 2011. http://www.ecochem.com/t_micronutrients.html.

Ellis, J.A., A.D. Walter, J.F. Tooker, M.D. Ginzel, P.F. Reag. "Conservation biological control in urban landscapes:Manipulating parasitoids of bagworm (Lepidoptera: Psychidae)." Biological Control, 2005: 99 - 107.

Evans, Erv. "Plants That Attract Beneficial Insects." North Carolina State University. http://www.ces.ncsu.edu/depts/hort/consumer/quickref/pest%20management/plants_attract_beneficial.html

Finn, E. M. "Featured Creatures: Robber Flies." University of Florida Institute of Food and Agricultural Sciences. 2009. http://entnemdept.ufl.edu/creatures/beneficial/flies/robber_flies.htm

Flint, Mary Louise and Steve Dreistadt. Natural Enemies Handbook: The Illustrated Guide to Biological Pest Control. Berkeley, Calif.: University of California Press, 1999.

Flint, M. L. "University of California Agriculture and Natural Resources." Pests in Gardens or Landscapes: Earwigs. 09 2002. http://www.ipm.ucdavis.edu/PMG/PESTNOTES/pn74102.html (accessed March 2011).

Follett, R.H. and D.G. Westfall. "Zinc and Iron Deficiencies." Colorado State University Extension. 2004. http://www.ext.colostate.edu/pubs/crops/00545.html.

Frank, Steven & James R. Baker. "North Carolina State University: North Carolina Cooperative Extension." Ornamentals and Turf: Bagworms. http://www.ces.ncsu.edu/depts/ent/notes/O&T/trees/ort081e/ort081e.htm (accessed February 17, 2011).

Genetzky, A, E. C. Burkness and W. D. Hutchison. "VegEdge: University of Minnesota." Four Lined Plant Bug. 2009. http://www.vegedge.umn.edu/vegpest/CUCS/4line.htm (accessed March 2011).

Gonsalves, Andre K. and Stephen A. Ferrei. Fusarium oxysporum. http://www.extento.hawaii.edu/kbase/crop/type/f_oxys.htm

Grubinger, Vern. "University of Vermont Extension." Flea Beetle Management. http://www.uvm.edu/vtvegandberry/factsheets/fleabeetle.html.

Hahn, Jeffrey. "University of Minnesota Extension." Fourlined Plant Bug. http://www.extension.umn.edu/yardandgarden/ygbriefs/e121plantbugs-fourlined.html.

Hahn, Suzanne Wold-Burkness and Jeffrey. "Asparagus Beetles in Home Gardens." University of Minnesota Extension. 2011. http://www.extension.umn.edu/distribution/horticulture/M1199.html (accessed February 15, 2011).

Hansen, Mary Ann. "Septoria Leaf Spot of Tomato." Virginia Tech: Virginia Cooperative Extension Service. 2009. http://pubs.ext.vt.edu/450/450-711/450-711.html.

Hartman, J. R., M. L. Witt, K. W. Wells, and W. O. Thom. Iron Deficiency of Landscape Plants. 1991. http://www.ca.uky.edu/agc/pubs/id/id84/id84.htm.

Hawkinson, Candice. "Galveston County Master Gardeners: Hover/Syphrid/Flower Flies." Texas A & M University: Extension Horticulture. 2006. http://aggie-horticulture.tamu.edu/galveston/beneficials/beneficial-28_hover_or_syrphid_flies.htm.

Hazelrigg, Ann. "Winter Injury." University of Vermont Extension. 1992. http://www.uvm.edu/pss/ppp/pubs/gd20.htm.

Hunter, Beatrice Trum. Gardening Without Poisons. Boston: Houghton Mifflin Company. 1964.

Imbriani, Dr. Jack L. Nematologist, NC Department of Agriculture (retired) (2011).

Jacobi, W. R. "Aspen and Poplar Leaf Spots." Colorado State University Extension. 2009. http://www.ext.colostate.edu/pubs/garden/02920.html.

Jones, Ronald K. and Charles W. Averre. "Damping-off in Flower and Vegetable Seedlings." North Carolina State University College of Agriculture and Life Sciences. 2000. http://www.ces.ncsu.edu/depts/pp/notes/oldnotes/od14.htm.

Kuepper, George. "Colorado Potato Beetle: Organic Control Options." National Sustainable Agriculture Information Service. 2003. http://attra.ncat.org/attra-pub/coloradopotato.html (accessed February 17, 2011).

Kuepper, George. "National Sustainable Agriculture Information Systen." Flea Beetle: Organic Control Options. 2003. http://attra.ncat.org/attra-pub/flea-beetle.html (accessed March 2011).

Lindemann, W.C. and C. R. Glover. Nitrogen Fixation by Legumes. New Mexico State University Cooperative Extension Service, 2003.

Linsley, Diane. Diane's Flower Seeds: Flowers for Beneficial Insects. 2007. http://www.dianeseeds.com/flowers/beneficial-insects.html.

Loomis, J. and H. Stone. "Lady Beetle Hippodamia convergens." Oregon State University Extension. April 2007. http://extension.oregonstatc.cdu/catalog/pdf/ec/ec1604.pdf

Loomis, J.M. and H. Stone. "Praying Mantis Stagmomantis californica." Oregon State University. April 2007.

Lyon, Williams F. "Ohio State University Extension Fact Sheet." Lady Beetle. http://ohioline.osu.edu/hyg-fact/2000/2002.html.

Mahr, Susan and Dan. "Tachinid Flies." Midwest Biological Control. http://www.entomology.wisc.edu/mbcn/kyf409.html

Mahr, Susan. "Aphidius Wasps." University of Wisconsin at Madison: Know Your Friends. http://www.entomology.wisc.edu/mbcn/kyf502.html.

Mahr, Susan. "Know Your Friends: Damsel Bugs." University of Wisconsin: Midwest Biological News. http://www.entomology.wisc.edu/mbcn/kyf402.html

McCain, Arthus H. and Robert D. Raabe and Stephen Wilhelm. Plants Resistant or Susceptible to Verticillium Wilt. Berkley: University of California, 1981.

McKenry, M. V. and P. A. Roberts. "Foliar (Bud and Leaf) Nematodes." University of California: Nematode Online Study Guice. 1985. http://groups.ucanr.org/nosg/Chapter_1/Review_Foliar_Nematodes.htm.

McPartland, J. M., R. C. Clarke and D. P. Watson. Hemp Diseases and Pests: Management and Biological Control. Cambridge: CAB International, 2000.

Mercure, Pamela S. "Damping off." University of Connecticut Integrated Pest Management. 1998. http://www.hort.uconn.edu/ipm/greenhs/htms/dampofgh.htm.

Meyer, John R. "Entomology 425 Compendium." Dermaptera. 2005. http://www.cals.ncsu.edu/course/ent425/compendium/earwigs.html#facts (accessed March 2011).

Meyerdirk, D. E., R. Warkentin, B. Attavian, E. Gersabeck, A. Biological Control of Pink Hibiscus Mealybug Project Manual. Manual, US Department of Agriculture, 2001.

Newton, Black. "Assassin bugs and Ambush bugs." University of Kentucky Department of Entomology.

2006. http://www.uky.edu/Ag/CritterFiles/casefile/insects/bugs/assassin/assassin.htm.

Newton, Black. "University of Kentucky Entomology: Kentucky Critter Files: Kentucky Insects." Kentucky Lacewings. 2008. http://www.uky.edu/Ag/Critter-Files/casefile/insects/lacewings/lacewings.htm

Newton, Blake. "Ground Beetles." University of Kentucky Department of Entomology. 2004. http://www.uky.edu/Ag/CritterFiles/casefile/insects/beetles/ground/ground.htm.

North Carolina State University: Center for Integrated Pest Management. AG-295: Insect and Related Pests of Vegetables: Mexican Bean Beetle. http://ipm.ncsu.edu/ag295/html/mexican_bean_beetle.htm.

Nunez, Joel. "Iron Deficiency in Vegetables ." University of Carlfornia Cooperative Extension. 2000. http://cekern.ucdavis.edu/Custom_Program573/Iron_Deficiency_in_Vegetables.htm.

Olkowski, William, Sheila Daar and Helga Olkowski, Common-Sense Pest Control: Least-Toxic Solutions for Your Home, Garden, Pets, and Community. Newton, Conn.: Taunton, 1991.

Olsen, Mary W. Damping Off. Tucson, 2011.

Oregon State University. "Damsel Bugs." IPMP3.0. http://uspest.org/mint/damselid.htm.

Pasteuria Bioscience. Product Development. 2009. http://pasteuriabio.net/product_development.html.

Perry, Leonard. "The Green Mountain Gardener: Salt Damage to Plants." University of Vermont Extension. http://www.uvm.edu/pss/ppp/articles/salt1.htm.

Pfeiffer, D. G. and H. W. Hogmire. "Aphid Predators." In Mid-Atlantic Orchard Monitoring Guide. New York: NRAES.

Plant Disease and Insect Diagnostic Clinic. "Septoria Leaf Spot ." Iowa State University. 2008. http://www.isuplantdiseaseclinic.org/node/120/default.htm.

Platnik, Norman I. The World Spider Catalog. 2011. http://research.amnh.org/iz/spiders/catalog/COUNTS.html

Potter, Michael F., Ric Bessin, and Lee Townsend. "ASIAN LADY BEETLE INFESTATION OF STRUCTURES." The University of Kentucky College of Agriculture. June 2005. http://www.ca.uky.edu/entomology/entfacts/ef416.asp.

Ragsdale, D. and E. Radcliffe. "Colorado Potato Beetle: VegEdge." Department of Entomology, University of Minnesota. June 2010. http://www.vegedge.umn.edu/vegpest/cpb.htm (accessed March 2011).

Rhode Island Landscape Horticulture Program. Lace Bugs. 1999. http://www.uri.edu/ce/factsheets/sheets/lacebugs.html (accessed March 2011).

Rincon-Vitova Insectaries, Inc. Aphidius: Aphid Parasites. Ventura, CA, 2006.

Roe, Alan H. Ground Beetles. Logan, Utah, 1989.

Rosenthal, Ed and Kathy Imbriani. Marijuana Pest and Disease Control: How to Protect Your Plants and Win Back Your Garden. Oakland, Calif.: Quick Trading Co. 2012

Sasser, J. N. and C. C. Carter. An Advanced Treatise on Meloidogyne . Raleigh: North Carolina State University Graphics, 1985.

Scheuerell SJ, Mahaffee WF. "Compost Tea as a Container Medium Drench for Suppressing Seedling Damping-Off Caused by Pythium ultimum." Phytopathology, Nov. 2004: 1156-63.

Sherf, Arden. "Verticillium wilt of Tomato." Cornell University . http://vegetablemdonline.ppath.cornell.edu/factsheets/Tomato_Verticillium.htm.

Shober, Amy L. and Geoffrey C. Denny. "Soil pH and the Home Landscape or Garden." University of Florida IFAS Extension. 2008. http://edis.ifas.ufl.edu/ss480.

Sikora, Edward J. "Ozone Damage to Plants." Alabama Cooperative Extension System. 2004. http://docs.google.com/viewer?a=v&q=cache:jYg1i2vrnVwJ:www.aces.edu/pubs/docs/A/ANR-0940/ANR-0940.pdf+ozone+damage+plants&hl=en&gl=us&pid=bl&srcid=ADGEESgCj6BeE3AZK_Km3QlOfeWfiysiTKiz5H-SEpz2_haZYwDqT0FvDhXcVdkHpBYBHCh6Ep-140Wa-dLUGzI-Bj6kepM8DcKjNlXK121_p5

Stitch, J. C. Marijuana Garden Saver. Oakland, California: Quick Trading Co., 2008.

Stokes, Donald W. A Guide to Observing Insect Lives. Boston: Little, Brown and Company, 1983.

Stuart, John. Mother Earth News: Tachinid Flies. 2004.

http://www.motherearthnews.com/Nature-Community/2004-08-01/Tachinid-Flies.aspx?page=1

Summer Hill Nursery. The Plants We Grow: A few thoughts about deer. http://www.summerhillnursery.com/plantswegrow/plants-deer.html.

Syngenta BioLine. Aphid control: Alphaline c.

Texas A & M System. "Praying Mantid." AgriLife Extension.

Texas A & M System. "Robbery Fly." AgriLife Extension.http://insects.tamu.edu/fieldguide/cimg228.html

The Beneficial Insect Company. Green Lacewing Larvae. 2010. http://www.thebeneficialinsectco.com/green-lacewing-larvae.htm

The University of Arizona: The Center for Insect Science Education Outreach . Praying Mantid Information. 1997. http://insected.arizona.edu/mantidinfo.htm

United Wildlife Control. Gopher Control. http://www.unitedwildlife.com/animalsgophers.html#c.

University of California Agriculture and Natural Resources. "Asparagus: Asparagus Beetles." University of California Agriculture and Natural Resources. June 2009. http://www.ipm.ucdavis.edu/PMG/r7300511.html (accessed February 15, 2011).

University of California Agriculture and Natural Resources. "Citrus Mealybugs." UC IPM ONline: State Integrated Pest Management Program. September 2008. http://www.ipm.ucdavis.edu/PMG/r10 .html (accessed March 2011).

University of California Agriculture and Natural Resources. "Cole Crops: Imported Cabbageworm." How to Manage Pests. October 2010. http://www.ipm.ucdavis.edu/PMG/r108301111.html (accessed March 2011).

University of California Agriculture and Natural Resources. "Snails and Slugs ." Statewide Integrated Pest Management Program. November 2009. http://www.ipm.ucdavis.edu/PMG/PESTNOTES/pn7427.html (accessed March 2011).

University of California Agriculture and Natural Resources. "Syrphid, flower, or hover flies." University of California: Statewide Integrated Pest Management Program. 2003. http://www.ipm.ucdavis.edu/PMG/NE/syrphid_flies.html.

University of California Agriculture and Natural Resources. "Thrips." How to Manage Pests. 2007. http://www.ipm.ucdavis.edu/PMG/PESTNOTES/pn7429.html.

University of California Agriculture and Natural Resources. "UC IPM Online Statewide Integrated Pest Management Program." Squash Bugs. 12 2009. http://www.ipm.ucdavis.edu/PMG/r116301111.html (accessed March 2011).

University of California. "Pythium Root Rot." University of California Agricultural and Natural Resources. 2009. http://www.ipm.ucdavis.edu/PMG/r280100211.html.

University of California. "Tomato Fusarium Wilt." University of California Agriculture and Natural Resources. 2007. http://www.ipm.ucdavis.edu/PMG/r783101011.html.

University of Florida Entomology Department. "Scale Insects." Woodybug Insects and Mites. http://entnemdept.ufl.edu/fasulo/woodypest/scales.htm.

University of Florida Institute of Food and Agricultural Services. "Big-eyed bugs ." Featured Creatures. 2004. http://entnemdept.ufl.edu/creatures/beneficial/bigeyed_bugs.htm#dist.

University of Florida Institute of Food and AgriculturalSciences. "Tobacco Budworm." Featured Creatures. July 2001. http://entnemdept.ufl.edu/creatures/field/tobacco_budworm.htm (accessed March 2011).

University of Illinois Extension. "Foliar Nematode Disease of Ornamentals." Extension Publication, Urbana - Champaign, 2000.

University of Illinois Extension. "Botrytis Blight or Gray Mold of Ornamental Plants." Report on Plant Disease, Urbana - Champaign: University of Illinois, 1997.

University of Rhode Island Landscape Horticulture Program. "Cutworms." Green Share Fact Sheets. 1999. http://www.uri.edu/ce/factsheets/sheets/cutworms.html (accessed March 2011).

University of Rhode Island Landscape Horticulture Program. "Imported Cabbageworm." Green Share Fact Sheets. 1999. http://www.uri.edu/ce/factsheets/sheets/importcabbageworm.html (accessed March 2011).

University of Rhode Island Landscape Horticulture

Program. "Pine Sawflies." Face Sheets. 1999. http://www.uri.edu/ce/factsheets/sheets/pinesawfly.html (accessed March 2011).

U.S. Department of Agriculture: Animal and Plant Health Inspection Service. Managing the Japanese Beetle: A Homeowner's Handbooik. April 2004.

Vann, Stephen, T. L. Kirkpatrick and Rick Cartwright. "Control Root-Knot Nematodes in Your Garden." University of Arkansas. www.uaex.edu/Other_Areas/publications/PDF/FSA-7529.pdf.

Vitosh, M.L., D.D. Warncke and R.E. Lucas. "Soils and Soils Management: Boron Deficiency." Michigan State University Extension . 1994. http://web1.msue.msu.edu/imp/modf1/05209709.htm

Wainwright-Evans, Suzanne. "Identifying the Enemy - Fungus Gnats." Interiorscape, 2002: 108.

Wallis, Chris and Dennis J. Lewandowski. "Fact Sheet: Black Spot of Roses." Ohio State Unversity Extension. 2008. http://docs.google.com/viewer?a=v&q=cache:mepm23xtXWoJ:ohioline.osu.edu/hyg-fact/3000/pdf/3072.pdf+black+spot+roses&hl=en&gl=us&pid=bl&srcid=ADGEESh3vD614eWd_hO-hJY1Lhx30uIzsQu83iGTyS5SSLx8n8XyruAj0afGh-nB2xRqu9RSpFdkdvjDzgbtLjnpCdE1v_MCObkd-sxWlmj9bhFOuSt5-.

Walliser, Jessica. Good Bug, Bad Bug. Pittsburgh, PA: St. Lynn's Press, 2008.

Washington State University. "Gardening in Western Washington." http://gardening.wsu.edu/library/comm001/comm001.htm.

Watkins, Gary and Ric Bessin. "Praying Mantids." University of Kentucky College of Agriculture. 2003. http://www.ca.uky.edu/entomology/entfacts/ef418.asp

Wawrzynski, Robert P. "University of Minnesota Extension." Sawflies of Trees and Shrubs. 2011. http://www.extension.umn.edu/distribution/horticulture/DG6703.html (accessed March 2011).

Wilson, Suzanne. "Deer Gardening." MDCOnline. http://mdc.mo.gov/conmag/2003/04/deer-gardening?page=0,0.

Wold-Burkness, S.J. & W.D. Hutchison. "Tomato Hornworm : VegEdge." University of Minnesota . March 2010. http://www.vegedge.umn.edu/vegpest/hornworm.htm (accessed March 2011).

Wright, Bob. "Minute Pirate Bugs." Midwest Biological Control News. http://www.entomology.wisc.edu/mbcn/kyf101.html

Yepsen, Roger B. The Encyclopedia of Natural Insect and Disease Control. Emmaus, Penn.: Rodale. 1984.

Ye, Weimin. "Root Knot Nematodes." North Carolina Department of Agriculture: Nematode Assay Section. 2008. http://www.ncagr.gov/agronomi/pubs.htm.

Zitter, T. A. and M. T. Banik. "Vegetable Crops: Virus Diseases of Cucurbits." Cornell University: Vegetable MD Online. 10 1984. http://vegetablemdonline.ppath.cornell.edu/factsheets/Viruses_Cucurbits.htm (accessed February 2011).

PHOTO CREDITS

wood.org; Western Flower Thrip: Jack T. Reed, Mississippi State University, Bugwood.org, **WHITEFLIES:** pp. 110-112, Silverleaf whitefly: Jeffrey W. Lotz, Florida Department of Agriculture and Consumer Services, Bugwood. org; Whitefly damage: David B. Langston, University of Georgia, Bugwood.org; Whiteflies: Tom Murray; **DISEASES–BLACK SPOT:** pp. 114-115, Black spot on rose: Margery Daughtry; Typical infection: Wikimedia public domain; **DAMPING OFF:** p. 116-118, Stem rot: Andrej Kunca, National Forest Centre–Slovakia, Bugwood.org; Stem rot: Martin Draper, USDA-NIFA, Bugwood.org; **GRAY & BROWN MOLD:** pp. 119-120, Botrytis on peony leaves: Marie Iannotti; *Botrytis cinerea* on ripe strawberry, and Strawberry rachis engulfed by gray mold: Scott Bauer, USDA Agricultural Research Service, Bugwood.org; Onion bulbs with symptoms Botrytis: Howard F. Schwartz, Colorado State University, Bugwood.org; Botrytis blight of Celosia: Department of Plant Pathology Archive, North Carolina State University, Bugwood.org; **POWDERY MILDEW:** pp. 122-124, Powdery mildew on squash leaves: Ed Rosenthal; Powdery Mildew on Sweet Pea: Jane Klein; Powdery Mildew on hydrangea, on Rose leaves, and on Lilac: Margery Daughtry; **ROOT DISEASES–*FUSARIUM* WILT:** pp. 126-128, *Fusarium* wilt on sugar beet: Howard F. Schwartz, Colorado State University, Bugwood.org; Common hop plant with *Fusarium* canker: David Gent, USDA Agricultural Research Service, Bugwood.org; *Fusarium* wilt on mum root: Margery Daughtry; *Fusarium* wilt symptoms on dry bean plants, and on Onion plants: Howard F. Schwartz, Colorado State University, Bugwood.org; **PHYTOPHTHORA:** pp. 129-130, Zucchini squash plant with *Phytophthora* wilt: Howard F. Schwartz, Colorado State University, Bugwood.org; *Phytophthora* on rhododendrum: Margery Daughtry; Pepper plant infected with *Phytophthora*: Paul Bachi, University of Kentucky Research and Education Center, Bugwood.org; Cabbage decay: Howard F. Schwartz, Colorado State University, Bugwood.org; **PYTHIUM:** pp. 132-134, Drybean seedlings showing symptoms of *Pythium* damping off: Mary Ann Hansen, Virginia Polytechnic Institute and State University, Bugwood.org; **RHIZOCTONIA:** pp. 134-135, Root rot/damping off symptoms: Clemson University–USDA Cooperative Extension Slide Series, Bugwood.org; Rhizoctonia on pepper stem: Gerald Holmes, Valent USA Corporation, Bugwood.org; **VERTICILLIUM WILT:** pp. 136-138, Hop plants with *Verticillium* wilt: David Gent, USDA Agricultural Research Service, Bugwood.org; Strawberry infested with *Verticillium* wilt: USDA, Brian Prechtel; Sunflower plants with *Verticillium* wilt: Howard F. Schwartz, Colorado State University, Bugwood.org; **SEPTORIA LEAF SPOT:** pp. 138-140, Septoria leaf spot on tomato leaf: William M. Brown Jr., Bugwood.org; Septoria leaf spot on celery leaf: Wikimedia; Septoria on

dry pea: Mary Burrows, Montana State University, Bugwood.org; **TOBACCO MOSAIC VIRUS:** pp. 140-141, Tobacco Mosaic Virus on petunia leaf: Margery Daughtry; Tobacco Mosaic Virus on Euphorbia: Frank Vincentz; **NUTRIENTS—BORON:** pp. 145-146, Early deficiency in tomato leaf: courtesy of Dr. Emanuel Epstein; Deficiency in apples and other fruits and vegetables: Mary Ann Hansen, Virginia Polytechnic Institute and State University, Bugwood.org; **CALCIUM:** pp. 146-178, Calcium deficiency in tomato: www.gardenersedge.com. Lettuce necrosis that spreads to older leaves: Gerald Holmes, Valent USA Corporation, Bugwood.org; **COPPER:** pp. 148-150, Copper deficient symptoms in tomato: courtesy of gardencentreguide.co.uk; Typical copper deficiency: courtesy of gardencentreguide.co.uk; Early signs of copper deficiency in corn: courtesy http://www.extension.uidaho.edu; **IRON:** p. 150, Severe deficiency in maple leaf: William M. Ciesla, Forest Health Management International, Bugwood.org; Iron deficiency in tomato leaf: courtesy of Dr. Emanuel Epstein; Iron chlorosis in citrus: Ed Rosenthal; **MAGNESIUM:** pp. 152, Cucumber leaf with deficiency, courtesy of The International Plant Nutrition Institute (IPNI); Magnesium deficiency in Phoenix palm: courtesy of www.fairchildgarden.org; **MANGANESE:** pp. 153-154, Manganese deficiency in tomato: Courtesy of Dr. Emanuel Epstein; Inernodal chlorosis in cucumber leaf: Nigel Cattlin/Photo Researchers, Inc.; **MOLYBDENUM:** pp. 155-156, Deficient tomato: courtesy of Dr. Emanuel Epstein; Leaves of sweet potato: courtesy of www.lucidcentral.org; **NITROGEN:** pp. 156-158, Tomato leaf: courtesy of Dr. Emanuel Epstein; Broccoli plant: Nigel Cattlin/Photo Researchers, Inc.; **PHOSPHORUS:** pp. 159-160, Cabbage leaves: Courtesy of *Mineral Deficiencies in Plants*; Tomato plant with phosporus deficiency: courtesy of Dr. Emanuel Epstein; **POTASSIUM:** pp. 161-162, Tomato leaf with potassium deficiency: courtesy of Dr. Emanuel Epstein; Ginger plant with deficiency: courtesy of The International Plant Nutrition Institute (IPNI); **SULFUR:** p. 154, Tomato with sulfur deficiency: courtesy of Dr. Emanuel Epstein; Corn plant with sulfur deficiency, courtesy www.extension.uidaho.edu; **ZINC:** pp. 165-166, Tomato with zinc deficiency: courtesy of Dr. Emanuel Epstein; Zinc deficient onions: courtesy of The International Plant Nutrition Institute (IPNI); **ENVIRONMENTAL STRESSES:** Illustrations: Hera Lee; p. 175, rootbound plant, courtesy of Smart Pots, p. 181, pH Down: Angela Bacca; p. 182, Hanna combo tester: courtesy of Hanna Instruments; p. 187, Winter injury on tomato: Paul Bachi, University of Kentucky Research and Education Center, Bugwood.org; **BENEFICIAL BIOLOGICALS:** p. 214, Aphid Midges: Brian-Valentine; Spined assassin bug: Mark Plonksy; p. 216, Big-Eyed bug: Wikimedia; Damsel Bugs: Richard Bartz, Munich, and Tom Murray; Ground Beetle: Jon Richfield; p. 219, Flower

Flies: Russ Ottens, University of Georgia, Bugwood.org, David Cappaert, Michigan State University, Bugwood.org, Charles Ray, Auburn University, Bugwood.org; Flower fly larva: Russ Ottens, University of Georgia, Bugwood.org; p. 220 Lacewing: Mark Plonsky; Lacewing, and Lacewing eggs: Tom Murray; p. 221, Lady Beetles: Brian Valentine; "The Argument" ant/ladybug: Mark Plonsky; p. 222, Mantid: Wikimedia-Alvesgaspar; Minute Pirate Bug: Tom Murray; p. 225, Predatory Mite: Stephen Luk; p. 227, Robber Fly struggling with Yellow Jacket: David Adams, Balancedrock Photography, Bugwood.org; Robber fly: courtesy of Community Gardens, Skidaway Island, GA; p. 228, Tachinid fly: Ron Napp; p. 229, Braconid Wasp Larva parasitizing caterpillar: Lloyd Davidson; Braconid wasp young hatch inside the aphid: Nick Dimmock, University of Northampton, Bugwood.org; **RESOURCE GUIDE:** p. 239, Plums: Scott Bauer, USDA Agricultural Research Service.

MOST COMMON PROBLEM PHOTOS: Aphid cluster: Ed Rosenthal; Western spotted cucumber beetle:, Susan Ellis, USDA APHIS PPQ, Bugwood.org; Japanese Beetle skeletonizing European linden: Steven Katovich, USDA Forest Service, Bugwood.org; Tobacco budworm: Scott Bauer, USDA Agricultural Research Service, Bugwood.org; Tomato hornworm last stage larva: Whitney Cranshaw, Colorado State University, Bugwood.org; Adult earwig: Johnny N. Dell, Bugwood.org; Ragwort flea beetle: Eric Coombs, Oregon Department of Agriculture, Bugwood.org; Fungus gnat: Tom Murray; Sycamore lace bug: courtesy of insectoid.info; Sycamore lace bug: Jim Baker, North Carolina State University, Bugwood.org; Citrus mealybug: Whitney Cranshaw, Colorado State University, Bugwood.org; Mole: yellow sunshine photos; Spider mite webbing: ©Povarov, Dreamstime.com; Squash bugs on pumpkin: Whitney Cranshaw, Colorado State University, Bugwood.org; Western Flower Thrip: Jack T. Reed, Mississippi State University, Bugwood.org; Whiteflies: Tom Murray; Stem rot: Andrej Kunca, National Forest Centre–Slovakia, Bugwood.org; *Botrytis cinerea* on ripe strawberry: Scott Bauer, USDA Agricultural Research Service, Bugwood.org; Powdery Mildew on Lilac: Margery Daughtry; Tobacco Mosaic Virus on petunia leaf: Margery Daughtry; Iron deficiency in tomato leaf: courtesy of Dr. Emanuel Epstein; Calcium deficiency in tomato: www.gardenersedge. com Magnesium deficiency in Phoenix palm: courtesy of www.fairchildgarden.org; Tomato leaf with potassium deficiency: courtesy of Dr. Emanuel Epstein.

BACK COVER PHOTOS: Strawberry rachis engulfed by gray mold: Scott Bauer, USDA Agricultural Research Service, Bugwood.org; Rootbound plant, courtesy of Smart Pots; Calcium deficiency in tomato: www.gardenersedge.com.

RESOURCE GUIDE

We would like to thank the companies whose support and participation helped make this book possible. Their products make gardening easier and safer for you, your harvest, and the planet.

It's Organic*, It's Effective, **It's Safer® Brand.**

OMRI® Listed insecticides, fungicides and herbicides.

www. saferbrand.com

*: for use in Organic Gardening

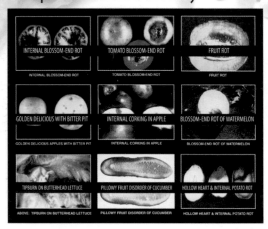

Organic Duo

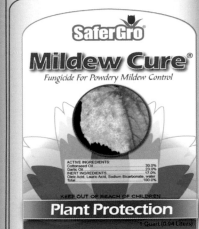
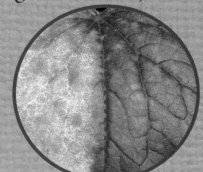
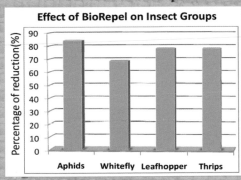

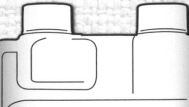

WEST COAST

HORTICULTURE

SPECIALTY ORGANIC FERTILIZERS

GROW 2-1-3 is an organic fertilizer made from fermented plant extract. It is formulated to provide ideal ratios of nitrogen, phosphorous and potassium as well as highly beneficial sugars, humic acid, organic acids and amino acids that support robust vegetative growth, roots and foliage. OMRI listed.

BLOOM 2-2-4 is an organic fertilizer made from fermented plant extract. It is formulated to provide ideal ratios of nitrogen, phosphorous and potassium as well as highly beneficial organic acids, humic acid and amino acids that support large, aromatic blooms. Naturally occurring sugars in this formula provide food for beneficial soil microbes. OMRI listed.

MICRONUTRIENTS is a specialty liquid formula that provides boron, copper, iron, manganese, molybdenum and zinc. Micronutrients are an essential part of a balanced nutrient feed program and affect aroma, flavor and yield. OMRI listed.

CHELATED CALCIUM is formulated with naturally occurring citric acid that aids in the absorption and improved transport of calcium throughout the plant. Calcium is necessary for cell division and is an essential building block for all stages of plant growth. Chelated Calcium is used to correct and prevent calcium deficiencies. OMRI listed.

MAGNESIUM is a liquid mineral formula that supplies a form of magnesium that is immediately available to the plant. Magnesium forms the central core of the chlorophyll molecule and activates enzymes necessary for plant growth and reproduction. Chlorophyll production is decreased when magnesium is deficient. This formula is used to correct and prevent Magnesium deficiencies. OMRI listed.

NATURAL PLANT GROWTH ENHANCER is used to promote robust vegetative growth, strong roots and foliage as well as larger and more aromatic blooms. It supplies a highly available form of potassium. This formula aids in the production and transportation of sugars and nutrients throughout the plant and helps enhance vegetative growth, bloom size and quality. OMRI listed.

ORGANIC BLOOM BOOSTER is used to enhance bloom size and quality. This product aids in the activation of enzymes and movement of chemical energy from leaves to flowers. It provides fully available phosphorous and potassium as well as sugars, amino acids and humic acid that help produce large and aromatic blooms.

MADE IN THE USA
www.westcoasthorticulture.com • 4110 SE Hawthorne Blvd., P.O. Box #701 Portland, OR 97214

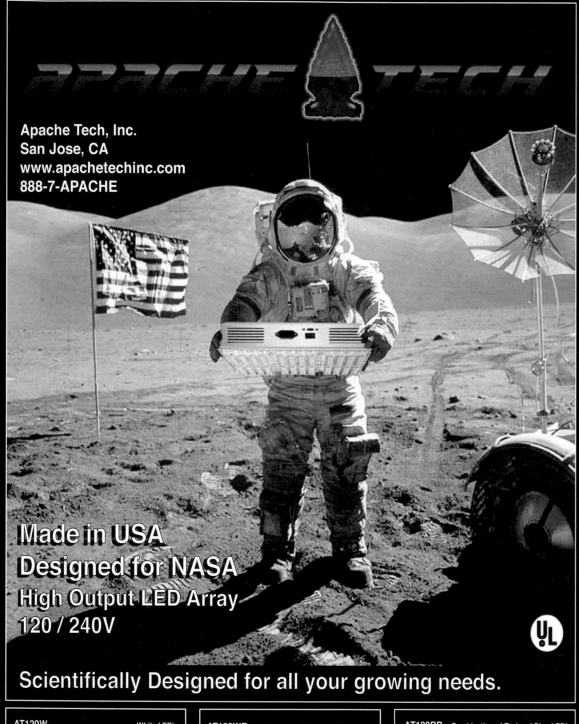
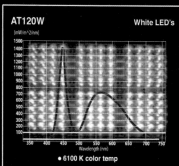
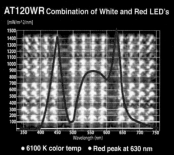
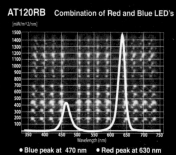

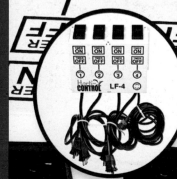

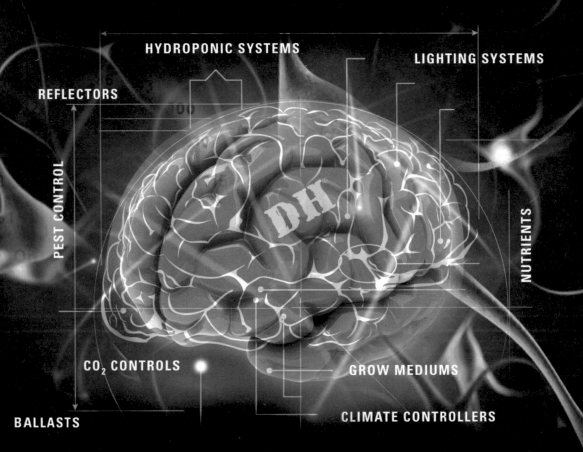